4th Workshop on Spatial Language Understanding and Grounded Communication for Robotics (SpLU-RoboNLP 2024)

Bangkok, Thailand
11-16 August 2024

ISBN: 979-8-3313-0178-1

Printed from e-media with permission by:

Curran Associates, Inc.
57 Morehouse Lane
Red Hook, NY 12571

proceedings.com

Some format issues inherent in the e-media version may also appear in this print version.

Association for Computational Linguistics
209 N. Eighth Street
Stroudsburg, Pennsylvania 18360

Phone: 1-570-476-8006
Fax: 1-570-476-0860

acl@aclweb.org

Additional copies of this publication are available from:

Curran Associates, Inc.
57 Morehouse Lane
Red Hook, NY 12571 USA
Phone: 845-758-0400
Fax: 845-758-2633
Email: curran@proceedings.com
Web: www.proceedings.com

SpLU-RoboNLP 2024

The 4th Workshop on Spatial Language Understanding and Grounded Communication for Robotics

Proceedings of the Workshop

August 11-16, 2024

Association for Computational Linguistics (ACL)
317 Sidney Baker St. S
Suite 400 - 134
Kerrville, TX 78028
USA
Tel: +1-855-225-1962
acl@aclweb.org

Introduction

Leveraging the foundation built in the prior workshops SPLU-RoboNLP-2022, SPLU-RoboNLP-2021, SpLU 2020, SpLU-RoboNLP 2019, SpLU 2018, and RoboNLP 2017, we organize the fourth combined workshop on Spatial Language Understanding and Grounded Communication for Robotics, SpLU-RoboNLP-2024. To achieve the long-term goal of natural conversation with robots in our homes, workplaces, hospitals, and warehouses, it is essential that we develop new techniques for linking language to perception and actions in the physical world.

This requires developing tools and theories to find insights into addressing some fundamental questions in NLP and HRI. Some important questions are the following. Can we give instructions to robotic agents to assist with navigation and manipulation tasks in remote settings? Can we talk to robots about the surrounding physical world, and help them interactively learn the language needed to finish a task? Can we develop robots that reply to us via grounded language generation, and eventually lead to an effective, two-way grounded dialogue? Given the rise of generative large language models, another question is how these large models can be deployed in situated dialogue settings and act meaningfully.

Human-robot dialogue often involves developing an understanding of grounded spatial descriptions. These capabilities invariably require understanding spatial semantics that relate to the physical environments where robots are embodied. Spatial semantics are the part of language semantics that is most related to grounding language into perception and the physical world. Spatial language meaning representation includes research related to cognitive and linguistically motivated spatial semantic representations, spatial knowledge representation and ontologies, qualitative and quantitative representation models, spatial annotation schemes, and efforts for creating specialized corpora. Spatial language learning considers both symbolic and sub-symbolic (with continuous representations) techniques and computational models for spatial information extraction, semantic parsing, and spatial co-reference within a global context that includes discourse and pragmatics from data or formal models. Recent studies show that one of the semantic aspects that pre-trained language models and even the recent large generative language models struggle with is reasoning over spatial language. We are interested in investigating whether qualitative and quantitative formal representations are helping spatial reasoning based on natural language and the possibility of learning such representations from data. Moreover, we emphasize on multimodality aspect of spatial language understanding as well as human-robot interaction. Some interesting related questions include, *which representations are appropriate for different modalities, and which ones are modality independent? How can we exploit visual information for language learning and reasoning?*

The main goal of this joint workshop is to bring in the perspectives of researchers working on physical robot systems and with human users and align spatial language understanding representation and learning approaches, datasets and benchmarks with the goals and constraints encountered in HRI and robotics. Such constraints include high costs of real-robot experiments, computational costs for real-time interactions, human-in-the-loop training and evaluation settings, scarcity of embodied data, as well as non-verbal communication.

The invited speakers, program committee, and organizing committee consist of researchers who belong to language, robotics, and vision communities or work in the intersection of these research areas.

We have 4 invited speakers, 3 archived papers, and several non-archival papers. Our workshop will accommodate the relevant ACL findings papers.

Organizing Committee

Organizing Committee

Parisa Kordjamshidi, Michigan State University
Xin Wang, University of California Santa Cruz
Yue Zhang, Michigan State University
Ziqiao Ma, University of Michigan
Mert Inan, Northeastern University

Program Committee

Program Committee

Cristian-Paul Bara, Amazon
Simon Dobnik, University of Gothenburg
Jiafei Duan, University of Washington
Yue Fan
Xiaofeng Gao, Amazon
Felix Gervits
Drew A. Hudson, Google DeepMind
Jacob Krantz, Facebook
Jialu Li, Department of Computer Science, University of North Carolina at Chapel Hill
Jiachen Li, University of California, Santa Barbara
Manling Li, Northwestern University
Weiyu Liu, Stanford University
Stephanie M. Lukin, DEVCOM Army Research Laboratory
Jiayi Pan, University of California, Berkeley
Natalie Parde, University of Illinois Chicago
Roma Patel, DeepMind and Brown University
Chris Paxton, meta
Yanyuan Qiao
Kirk Roberts, University of Texas Health Science Center at Houston
Raphael Schumann
Yichi Zhang, University of Michigan

Table of Contents

Language-guided World Models: A Model-based Approach to AI Control
Alex L Zhang, Khanh Xuan Nguyen, Jens Tuyls, Albert Lin and Karthik R Narasimhan 1

Learning Communication Policies for Different Follower Behaviors in a Collaborative Reference Game
Philipp Sadler, Sherzod Hakimov and David Schlangen 17

Collection of Japanese Route Information Reference Expressions Using Maps as Stimuli
Yoshiko Kawabata, Mai Omura, Hikari Konishi, Masayuki Asahara and Johane Takeuchi 30

Program

Friday, August 16, 2024

09:00 - 09:15 *Opening Remarks*

09:15 - 10:00 *Invited Talk1*

10:00 - 10:30 *Grounded Communication for Robotics*

Language-guided World Models: A Model-based Approach to AI Control
Alex L Zhang, Khanh Xuan Nguyen, Jens Tuyls, Albert Lin and Karthik R Narasimhan

Natural Language Can Facilitate Sim2Real Transfer
Albert Yu, Adeline Foote, Ray Mooney and Roberto Martín-Martín

10:30 - 11:00 *Coffee Break*

11:00 - 11:45 *Invited Talk2*

11:45 - 12:30 *Grounded Communication for Robotics*

Into the Unknown: Generating Geospatial Descriptions for New Environments
Tzuf Paz-Argaman, John Palowitch, Sayali Kulkarni, Reut Tsarfaty and Jason Michael Baldridge

Tuning Language Models with Spatial Logic for Complex Reasoning
Tanawan Premsri and Parisa Kordjamshidi

TopViewRS: Vision-Language Models as Top-View Spatial Reasoners
Chengzu Li, Caiqi Zhang, Han Zhou, Nigel Collier, Anna Korhonen and Ivan Vulić

12:30 - 14:00 *Lunch*

14:00 - 14:45 *Invited Talk3*

14:45 - 15:30 *Invited Talk4*

Friday, August 16, 2024 (continued)

15:30 - 16:00 *Coffee Break*

16:00 - 16:20 *Poster Spotlight*

16:20 - 17:30 *Posters*

Language-Guided World Models 
A Model-Based Approach to AI Control

[*]**Alex Zhang**[◇], [*]**Khanh Nguyen**[♠], **Jens Tuyls**[◇], **Albert Lin**[♣], **Karthik Narasimhan**[◇]
[◇] Princeton University [♠]University of California, Berkeley
[♣]University of Southern California

Project website: `language-guided-world-model.github.io`

Abstract

This paper introduces the concept of *Language-Guided World Models* (LWMs)—probabilistic models that can simulate environments by reading texts. Agents equipped with these models provide humans with more extensive and efficient control, allowing them to simultaneously alter agent behaviors in *multiple* tasks via natural verbal communication. In this work, we take initial steps in developing robust LWMs that can generalize to compositionally novel language descriptions. We design a challenging world modeling benchmark based on the game of MESSENGER (Hanjie et al., 2021), featuring evaluation settings that require varying degrees of compositional generalization. Our experiments reveal the lack of generalizability of the state-of-the-art Transformer model, as it offers marginal improvements in simulation quality over a no-text baseline. We devise a more robust model by fusing the Transformer with the EMMA attention mechanism (Hanjie et al., 2021). Our model substantially outperforms the Transformer and approaches the performance of a model with an oracle semantic parsing and grounding capability. To demonstrate the practicality of this model in improving AI safety and transparency, we simulate a scenario in which the model enables an agent to present plans to a human before execution, and to revise plans based on their language feedback.

1 Introduction

Model-based agents are artificial agents equipped with probabilistic "world models" that are capable of foreseeing the future state of an environment (Deisenroth and Rasmussen, 2011; Schmidhuber, 2015). World models endow these agents with the ability to plan and learn in imagination (i.e., internal simulation) and have led to exciting results in the field of reinforcement learning (Finn and Levine, 2017; Ha and Schmidhuber, 2018; Chua et al., 2018; Hafner et al., 2023). These models have been studied extensively for the purpose of improving the autonomous performance of artificial agents.

In this paper, we endorse and enhance the model-based approach for a different goal: to strengthen the controllability of artificial agents. Since all policies of a model-based agent are optimized with respect to a common world model, a human can adjust multiple policies simultaneously by making appropriate changes to this model. This mechanism complements the model-free approach that updates policies individually, offering greater efficiency and flexibility in control. For example, by incorporating the fact that the floor is slippery into the world model of a robot, a person can effectively remind it to handle *every* object in a room with greater caution. If the performance of the robot on a task remains unsatisfactory, the person can continue to fine-tune its policy for that specific task. In contrast, without a world model, they have to separately adapt the robot's policies to the slippery-floor condition.

The model-based approach requires world models that can be easily modulated by humans. Traditional world models fall short in this quality because they can only be modified using observational data, which is not a suitable medium for humans to convey intentions (Sumers et al., 2023; Zheng et al., 2023). To overcome the limitations of these models, we develop ***Language-Guided World Models*** (LWMs)—world models that can be effectively steered through human verbal communication. Agents equipped with LWMs inherit all the benefits of model-based agents while being able to incorporate language-based supervision. This capability reduces human teaching effort and mitigates the risk of agents taking harmful actions in an environment to explore its dynamics. LWM-based agents can also self-improve by reading "free" texts

[*]First two authors contribute equally. Correspondence email: `kxnguyen@berkeley.edu`.

Proceedings of the 4th Workshop on Spatial Language Understanding and Grounded Communication for Robotics (SpLU-RoboNLP 2024), pages 1–16

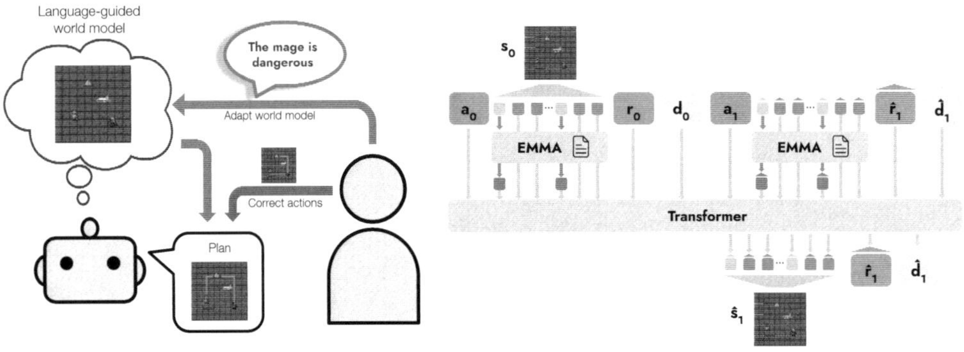

(a) (b)

Figure 1: Language-guided world models (LWMs) offer human an efficient mechanism to regulate artificial agents. (a) We illustrate a potential application of LWMs to improving AI safety and transparency. These models enable an agent to generate visual plans and invite a human supervisor to validate them. Moreover, the human can adjust the plans by modifying the agent's world model with language feedback, in addition to directly correcting its policy. (b) We design an architecture for LWMs that exhibits strong compositional generalization. We replace the cross-attention mechanism of the standard Transformer with a new attention mechanism inspired by Hanjie et al. (2021) to effectively incorporate language descriptions. We then train a model that auto-regressively generates tokenized observations conditioned on language descriptions and actions.

composed to guide humans (e.g., game manuals), reducing the subsequent effort to fine-tune them through direct interaction.

Building LWMs poses a unique research challenge: grounding language to environmental dynamics. This problem is difficult because the language used to describe environment dynamics can be incredibly rich and complex, encompassing a wide range of concepts such as entity names, appearances, motions, interactions, spatial and temporal relations, and more. Moreover, in natural settings, especially when describing artificial environments (e.g., games), new concepts are often introduced but may not always be clearly defined. Humans deal effectively with this issue because they possess remarkable reasoning capabilities that allow them to infer word meanings from observations. For example, a caption like "*the Ziff, which is chasing the player, is extremely hostile*" and a video depicting this scene likely provide enough clues for a person to determine what "the Ziff" refers to, assuming that they are familiar with the concept of "chasing". Not only understanding word meanings, humans are also capable of applying newly learned words in novel ways, enabling imagination of new dynamics, such as envisioning a "fleeing Ziff" that runs away from the player.

Toward building world models with similar capabilities, we construct a benchmark based on the game of MESSENGER (Hanjie et al., 2021). In this benchmark, a model is given trajectory "videos" of games involving several entities interacting with each other. Each video is accompanied by language descriptions of the attributes of the entities. The model begins with almost zero language understanding and has to identify the entities and learn the grounded meanings of their attributes purely by watching the videos. At test time, it must demonstrate *compositional generalization* by being able to simulate environments featuring entities with attributes different from those it observes during training. For example, it has to portray a "fleeing mage" despite having only seen the mage chase the player in training games. We design three evaluation settings that test for incrementally greater degree of compositional generalization.

Despite its apparent simplicity, our benchmark covers many complications in building robust LWMs. We find that the prominent Transformer model (Vaswani et al., 2017) struggles in the harder evaluation settings. Even with a ground-truth disentangled representation of the observations, the model cannot learn generalizable grounding functions and yields minimal improvements in simulation quality compared to a model that ignores the language descriptions entirely. We augment the model with the EMMA attention (Hanjie et al., 2021), which mimics a two-step reasoning process. Our results confirm the effectiveness of this new architecture, as it robustly generalizes even in the hardest evaluation setting, outperforming baselines by substantial margins in various evaluation metrics. It is even competitive with a skyline model with an oracle semantic parsing and grounding capability.

Last but not least, we illustrate a promising application of LWMs by simulating a cautious agent that, instead of performing a task right away, uses its LWM to generate an execution plan and asks a human to review it (Figure 1a). This form of pre-execution communication can potentially improve the agent's safety and transparency, following the spirit of the guaranteed safe AI approach proposed by Dalrymple et al. (2024). Moreover, it allows the human to improve the performance of the agent by revising the plan. In this setting, our LWM-based agent has the advantage of being able to assimilate *language feedback* describing the environment dynamics. We demonstrate that the language understanding capabilities of our proposed LWM are sufficient to enact this strategy. In the most challenging evaluation setting, without gathering additional interactions in the environment, the agent equipped with our model achieves an average reward three to four times higher than that of an agent using an observational world model.

We hope that our work will serve as a catalyst for exploring novel approaches to developing robust language-guided world models. More generally, we call for the design of modular agents whose components are parameterized by natural language. As previously argued, a modular design can dramatically boost communication efficiency, because the same component may be involved in the learning of various policies. We hypothesize that this approach can potentially surpass the efficiency of the currently prevalent approach that integrates language into a monolithic policy (e.g, Bisk et al. (2016); Misra et al. (2018); Anderson et al. (2018); Narasimhan et al. (2018); Hanjie et al. (2021); Zhong et al. (2021) and work on large language models like Ouyang et al. (2022)).

2 Background: world models

We consider a Markov Decision Process (MDP) environment E with state space $\mathcal{S}$, action space $\mathcal{A}$, and transition function $M : \mathcal{S} \times \mathcal{A} \to \Delta(\mathcal{S} \times \mathbb{R} \times \{0, 1\})$, where Δ denotes the set of all probability distributions over a set. An agent implementing a policy $\pi(a \mid s) : \mathcal{S} \to \Delta(\mathcal{A})$ interacts with the environment by choosing actions using its policy. Taking an action $a_t \sim \pi(s_t)$ in state s_t transitions the agent to a new state s_{t+1}, and incurs a reward r_{t+1} and a termination signal d_{t+1}, where $s_{t+1}, r_{t+1}, d_{t+1} \sim M(s_t, a_t)$.

A (one-step) *world model* M_θ (Robine et al., 2023; Micheli et al., 2023; Hafner et al., 2023) is an approximation of $M(s_{t+1}, r_{t+1}, d_{t+1} \mid s_t, a_t)$. A *model-based agent* uses data gathered in the environment to construct a world model and leverages it to learn policies for accomplishing tasks.[1] In contrast, a *model-free* agent learns its policies directly from data collected in the environment.

Model-based agents can require less effort to adapt. Because all policies of a model-based agent are derived from a shared world model, any modifications made to this model would affect all of them. This feature can be exploited to reduce human effort in controlling this type of agent. Specifically, suppose we concern m tasks in the environment, necessitating m policies. If there is a change in the environment dynamics, a model-based agent only needs task-agnostic data to replicate this change in its world model. It can then re-optimize its policies with respect to the updated model. Meanwhile, a model-free agent needs to collect task-specific data to re-train all of its m policies. The data collection cost of the model-free approach scales with m, whereas that of the model-based approach is independent of m, since the policy re-optimization step uses only data generated by the world model.

Observational world models. The dominant approach to world modeling learns a function $M_\theta(s_{t+1}, r_{t+1}, d_{t+1} \mid h_t)$ parameterized by a neural network θ and conditioned on a history $h_t = (s_1, r_1, d_1, a_1, \ldots, s_t, r_t, d_t, a_t)$. We refer to this class of models as *observational world models* because they can be adapted with only observational data, through either in-weight learning (updating the model parameters to fit a dataset of observations), or in-context learning (plugging in a history of observations).

Relying on observation-based adaptation leads to two drawbacks. First, controlling these models is difficult because observations are inadequate for conveying complex, abstract human intentions. Second, collecting observations requires taking real actions in the environment, which can be expensive, time-consuming, and risky.

[1] Note that M_θ includes a reward function but can be combined with any other reward function for learning.

3 Language-guided world models (LWMs)

We introduce LWMs, a new class of world models that can interpret language descriptions to simulate environment dynamics. These models address the drawbacks of observational world models. They allow humans to easily adapt their behavior through natural means of communication. Consequently, humans can effectively assist these models, significantly reducing the amount of interactive experiences that they need to collect in environments. In addition, these models can also leverage pre-existing texts written for humans, saving human effort to fine-tune them.

3.1 Formulation

We consider a family of environments $E(\boldsymbol{v})$ whose transition function has the form $M(s_{t+1}, r_{t+1}, d_{t+1} \mid h_t, \boldsymbol{v})$ where $\boldsymbol{v}$ is a parameter vector. Plugging in a specific $\boldsymbol{v}$ gives rise to an environment. We assume that each environment $E(\boldsymbol{v})$ is accompanied by a *language manual* $\boldsymbol{\ell} = (l_1, \cdots, l_N)$ consisting of language descriptions l_i. This manual describes $\boldsymbol{v}$ and the internal operations of M. Our goal is to learn a world model $M_\theta(s_{t+1}, r_{t+1}, d_{t+1} \mid h_t, \boldsymbol{\ell})$ that approximates the true dynamics $M(s_{t+1}, r_{t+1}, d_{t+1} \mid h_t, \boldsymbol{v})$.

The training data for our LWMs is a dataset $\{(\tau^i, \boldsymbol{\ell}^i)\}$ where τ^i is a trajectory generated in an environment $E(\boldsymbol{v}_i)$ with $\boldsymbol{v}_i$ drawn from some distribution P_{train}, and $\boldsymbol{\ell}^i$ is the accompanying manual. Each trajectory $\tau = (s_1, r_1, d_1, a_1, \ldots, s_T, r_T, d_T)$ is a sequence of states, actions, rewards, and termination signals. It can be viewed as a "video" that is annotated with actions and rewards. The trajectories are generated using a behavior policy, which can be a rule-based or learned policy, or a human.

3.2 Modeling entity-based environments

We view an environment as a set of C *entities* interacting with each other within a constrained space. Each entity c has a set of K *attributes*, each of which has value v_k^c. There is a special attribute called the identity of the entity (e.g., the name of a character or object in a video game). Each action triggers an event that changes a subset of attributes of a group of entities. The specific change is determined by the attributes of the entities involved in the event (e.g., an enemy entity attacks a player when colliding with them). In this work, we assume that each description in a manual portrays all attributes of an entity; hence, the number of descriptions N is equal to C.

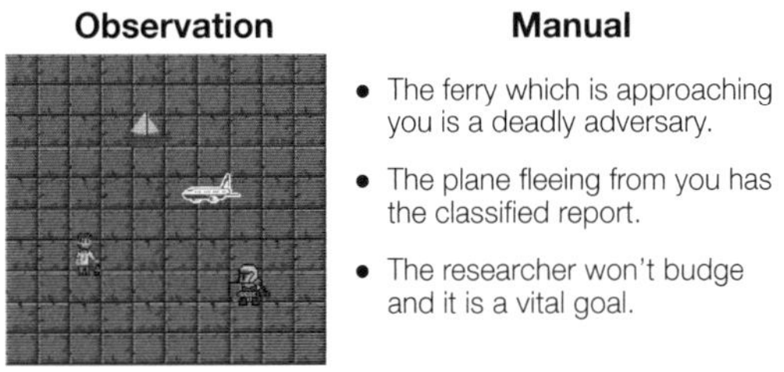

Figure 2: MESSENGER environment with manual.

Testing for compositional generalization. With this formulation, the environment parameters $\boldsymbol{v} = (v_1^1, \cdots, v_K^1, v_1^2, \cdots, v_1^C, \cdots, v_K^C)$ is a vector that contains the attributes of the C entities depicted in a manual. We are concerned with building LWMs that, at test time, can simulate environments whose paramerer vectors are compositionally novel. The term "compositionally novel" means that all components of the vector are individually seen during training, but the vector as a whole is previously unseen. This implies that the manuals at test time are also new.

This problem requires a LWM to be able to learn a representation of the transition function $M(\boldsymbol{v})$ by studying the language of the manuals, and to extract the specific parameters $\boldsymbol{v}$ described by each manual. The function $M(\boldsymbol{v})$ has two important properties. The first is the *independence* among its parameters because they represent orthogonal attributes. The second is the *locality* of the parameters, as each is an attribute associated with only a single entity. These properties make it difficult to recover the function exactly from purely observational data without injecting strong inductive biases into the learning model.

3.3 The MESSENGER-WM benchmark

The game of MESSENGER, developed by (Hanjie et al. (2021); Figure 2) exemplifies the class of environments discussed in the previous section. Despite being a simple grid-world environment, the dynamics possess the independence and locality properties that we want to study. In fact, it is our intention to use this visually simplistic environment to highlight the challenges in building LWMs that are orthogonal to the computer graphics challenge of mapping state representations to realistic-looking outputs.

Environment dynamics. The game takes place in a 10×10 grid world. A player interacts with entities of three *roles*: message, goal, and enemy. We use the stage-two version of the game, in which there are three entities, one of each role, in a game instance. In addition to the role, each entity is assigned an *identity* among twelve possibilities (mage, airplane, orb, etc.) and a *movement pattern* (chasing the agent, fleeing from the agent, immobile). The objective of the player is to acquire the message and deliver it to the goal while avoiding the enemy. Fetching the message is awarded 0.5 points and delivering it to the goal adds another point. If the player collides with the enemy or reaches the goal without carrying the message, the game ends, and the player receives -1 points.

Game manual. A game's manual consists of three descriptions corresponding to the three entities. MESSENGER provides a dataset of 5,316 language descriptions, each of which describes a combination of identity, role, and movement. The descriptions employ various linguistic expressions for each identity, role, or movement pattern (e.g., an airplane can be mentioned as a "plane", "jet", or "airliner"), making it non-trivial to interpret.

Evaluation settings. To test for compositional generalization, we construct three evaluation settings, ordered in increasing degree of difficulty:

- **NewCombo (easy).** Each game features a combination of three identities that were never seen together in a training game. However, the role and movement pattern of each identity are the same as during training.
- **NewAttr (medium).** The three identities were seen together in a training game, but each identity is assigned at least a new attribute (role, or movement pattern, or both).
- **NewAll (hard).** This setting combines the difficulties of the previous two. The identity triplet is novel, and each identity is assigned at least a new attribute.

To generate trajectories, we implement rule-based behavior policies that execute various intentions: act randomly, avoid the enemy, suicide (go to the enemy), obtain the message, and win the game (obtain the message and deliver it to the goal). We generate a total of 100K trajectories for training, each of which is generated by rolling out a uniformly randomly chosen rule-based policy. More details of the data are given in Appendix B. Our evaluation is more comprehensive than the original MESSENGER paper's evaluation, which does not construct different levels of compositional generalization, and is more difficult than the setting of Lin et al. (2024), which does not concern generalization.

To succeed in MESSENGER-WM, a model must be able to understand the non-trivial concepts represented by the attributes. For example, the concept of "chasing" involves planning actions to reduce the distance between two entities. The model must also capture the independence of the attributes, despite observing correlations in the training data (e.g., the "mage" is never immobile during training). Finally, to reflect the locality of the attributes, the model needs to learn a representation that disentangles the entities and to route attributes to the right entities. For example, the movement of one entity should not influence that of another. These are among the difficult, under-explored problems in machine learning, making MESSENGER-WM a respectable research challenge. We will empirically show that the state-of-the-art Transformer architecture struggles to perform well on the benchmark, suggesting that it may be insufficient for tackling more complex world-modeling problems.

4 Modeling approach

State representation. In MESSENGER, a state s is represented by an $H \times W$ grid with C channels (an $H \times W \times C$ tensor), where each channel corresponds to an entity. In each channel c, there is a single non-zero cell $s(h, w, c)$ that represents the identity of the entity. The position of this cell is the location of the entity in the grid. We note that this is an idealized representation that disentangles the entities. Even so, the problem remains challenging, as the model needs to recognize attributes mentioned in the manual and associate them with the right entity token. This requires a special attention mechanism, which we will introduce shortly. Meanwhile, learning entity-disentangled representations for pixel-based environments remain an open problem, which we defer to future work.

World modeling as sequence generation. Our model (illustrated in Figure 1b) is an encoder-decoder Transformer (Vaswani et al., 2017) which encodes a manual $\boldsymbol{\ell}$ and decodes a trajectory τ. We transform the trajectory into a long sequence of tokens and train the model as a sequence generator.

Concretely, our model processes a data point $(\tau, \boldsymbol{\ell})$ as follows. For the manual $\boldsymbol{\ell} = \{l_i\}_{i=1}^{N}$, we first use a pre-trained BERT model to convert each description l_i into a sequence of hidden vectors. We feed each sequence to a Transformer encoder, which outputs a tensor $\boldsymbol{m}^{\text{enc}}$ of size $N \times L \times D$, where $N = C$ is the number of descriptions, L is the maximum number of words in a description, and D is the hidden size.

For the trajectory, we convert each tuple (a_{t-1}, s_t, r_t, d_t) into a token block B_t. The first action a_0 is set to be a special <s> token. Each state s_t is mapped to $3C$ tokens $(i_t^1, h_t^1, w_t^1, \cdots, i_t^C, h_t^C, w_t^C)$, which represents each of the C entities by its identity i followed by its location (h, w). The real-valued reward r_t is discretized into an integer label, and the termination signal d_t is translated into a binary label. In the end, B_t consists of $3C + 3$ tokens $(a_{t-1}, i_t^1, h_t^1, w_t^1, \cdots, i_t^C, h_t^C, w_t^C, r_t, d_t)$. Finally, we concatenate all T blocks in the trajectory into a sequence of $T \times (3C + 3)$ tokens, embed them into a $T \times (3C + 3) \times D$ tensor, and add positional embeddings. We will use bold notation (e.g., $\boldsymbol{a}$, $\boldsymbol{i}$) to refer to the resultant embeddings of the tokens.

Entity mapper with multi-modal attention. We implement a variant of EMMA (Hanjie et al. (2021)) that first identifies the description that mentions each entity and extracts from it words corresponding to the attributes of the entity. From the tensor $\boldsymbol{m}_n^{\text{enc}}$ computed by the encoder, we generate a key tensor $\boldsymbol{m}^{\text{key}}$ and a value tensor $\boldsymbol{m}^{\text{val}}$, both of which are of size $N \times L \times D$, where

$$\boldsymbol{m}_n^{\text{key}} = \texttt{Softmax}(\texttt{Linear}_{\text{key}}(\boldsymbol{m}_n^{\text{enc}})^\top)\boldsymbol{m}_n^{\text{enc}}$$
$$\boldsymbol{m}_n^{\text{val}} = \texttt{Softmax}(\texttt{Linear}_{\text{val}}(\boldsymbol{m}_n^{\text{enc}})^\top)\boldsymbol{m}_n^{\text{enc}} \quad (1)$$

for $1 \leq n \leq N$. Here, $\texttt{Linear}_{\text{key}}^{D\to 1}$ and $\texttt{Linear}_{\text{val}}^{D\to 1}$ are linear layers that transform the input's last dimension from D to 1, and $\texttt{Softmax}(\cdot)$ applies the softmax function to the last dimension. Intuitively, we want each $\boldsymbol{m}_n^{\text{key}}$ to retain words that signal the identity of the entity mentioned in the n-th description (e.g., *ferry*, *plane*, *researcher*), and $\boldsymbol{m}_n^{\text{val}}$ to retrieve words depicting the other attributes (e.g., *approaching*, *deadly*, *fleeing*).

Let $\boldsymbol{i}_t^c$ be the embedding of the identity of entity c. We perform a dot-product attention with $\boldsymbol{i}_t^c$ as the query, $\boldsymbol{m}^{\text{key}}$ as the set of keys, and $\boldsymbol{m}^{\text{val}}$ as the set of values to compute the attribute features of c

$$\boldsymbol{z}_t^c = \texttt{DotAttend}(\boldsymbol{i}_t^c, \boldsymbol{m}^{\text{key}}, \boldsymbol{m}^{\text{val}}) \quad (2)$$

The features are added to the identity tokens $\boldsymbol{i}_t^c$. The final input of the model is as follows:

$$(\boldsymbol{a}_{t-1}, (\boldsymbol{i}_t^c + \boldsymbol{z}_t^c, \boldsymbol{h}_t^c, \boldsymbol{w}_t^c)_{c=1}^C, \boldsymbol{r}_t, \boldsymbol{d}_t) \quad (3)$$

Unlike the standard encoder-decoder Transformer, our architecture does not perform cross-attention between the encoder and the decoder because information from the encoder has already been incorporated into the decoder through EMMA.

Model training. We train the model to minimize cross-entropy loss with respect to the ground-truth (tokenized) trajectories in the training set. The label at each output position is the next token in the ground-truth sequence. In particular, we do not compute the losses at the positions of the action tokens and the first block's tokens, because those tokens will be set during inference.

5 Experiments

5.1 Baselines

We compare our model, which we call `EMMA-LWM`, with the followings:

(a) **`Observational`** world model does not leverage textual information. It is identical to `EMMA-LWM` except that we zero out the manual representation $\boldsymbol{m}^{\text{enc}}$;
(b) **`Standard`** is the encoder-decoder Transformer model following Vaswani et al. (2017) with multi-headed cross-attention between the decoder and the encoder. Similarly to `EMMA-LWM`, the model uses BERT to initially encode the manual into hidden vectors. The encoder applies self-attention to the hidden vectors of each description separately, instead of joining all vectors into a sequence and applying self-attention to it;
(c) **`GPTHard`** is similar to `EMMA-LWM` but uses ChatGPT instead of EMMA to ground descriptions to entities. More details about this model are in Appendix A;
(d) **`OracleParse`** is the same as `GPTHard`, but uses an oracle information extraction function. A description like *"the crucial target is held by the wizard and the wizard is fleeing from you"* is converted into *"mage fleeing goal"* for this model.

We train all models using AdamW (Loshchilov and Hutter, 2017) for 10^5 iterations. For further details, please refer to Appendix C.

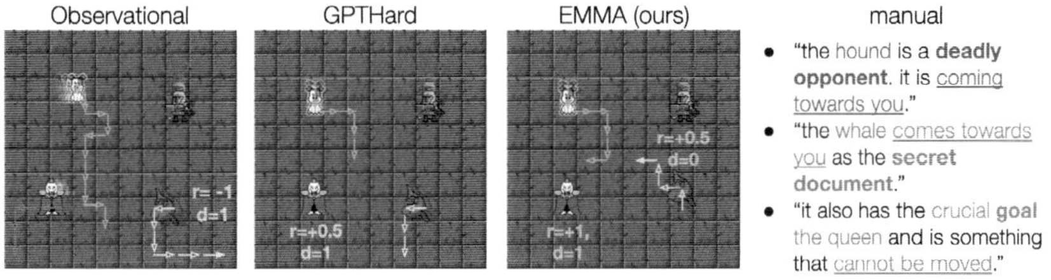

Figure 3: A qualitative example taken from the NewAll split. The `Observational` model mistakenly captures the movement patterns of the immobile queen goal and the chasing whale message. It also misrecognizes the whale as an enemy, predicting a wrong reward r and incorrectly predicting a termination state d after the player collides with this entity. The `GPTHard` model incorrectly identifies the queen as the message and predicts the whale to be fleeing. Meanwhile, our model `EMMA-LWM` accurately captures all of those roles and movements.

Table 1: Cross entropy losses (↓) of different models on test ground-truth trajectories. Note that the minimum loss is non-zero because the MESSENGER environment is stochastic. We run each model with five different random seeds, selecting the final checkpoint for each seed based on the loss in the development NewAll split. We report the mean losses with 95% t-value confidence intervals. The bold number in each column indicates the best non-oracle mean.

World model	NewCombo (easy)	NewAttr (medium)	NewAll (hard)
`Observational`	0.12 $\pm$ 0.04	0.18 $\pm$ 0.02	0.19 $\pm$ 0.01
`Standard`	0.10 $\pm$ 0.04	0.15 $\pm$ 0.04	0.16 $\pm$ 0.03
`GPTHard`	0.10 $\pm$ 0.02	0.15 $\pm$ 0.01	0.16 $\pm$ 0.00
`EMMA-LWM`	**0.08** $\pm$ 0.01	**0.10** $\pm$ 0.02	**0.13** $\pm$ 0.01
`OracleParse`	0.08 $\pm$ 0.01	0.09 $\pm$ 0.02	0.12 $\pm$ 0.06

5.2 Results

Evaluation with ground-truth trajectories. Table 1 shows the cross-entropy losses of all models on ground-truth trajectories sampled from the true environment dynamics (more in Appendix E). In the more difficult NewAttr and NewAll splits, our `EMMA-LWM` model consistently outperforms all baselines, nearing the performance of the `OracleParse` model. As expected, the `Observational` model is easily fooled by spurious correlations between identity and attributes, and among attributes. A specific example is illustrated in Figure 3. There, the `Observational` model incorrectly captures the movement of the whale and the queen. It also mistakenly portrays the whale as an enemy, whereas, in fact, the entity holds the message. In contrast, `EMMA-LWM` is capable of interpreting the previously unseen manual and accurately simulates the dynamics.

The performance of the `Standard` model is sensitive to initialization; in some runs, it performs as well as `EMMA-LWM`, but in others it performs as badly as `Observational`. A plausible explanation is that the model's attention mechanism lacks sufficiently strong inductive biases to consistently find generalizable solutions. Our results agree with previous work on the lack of compositional generalizability of Transformers, which is often remedied by adding various forms of inductive bias (Keysers et al., 2020; Jiang and Bansal, 2021; Chaabouni et al., 2021; Dziri et al., 2023).

Another interesting finding is that the `GPTHard` model does not perform as well as expected. As a reminder, this model relies on ChatGPT to parse identities from descriptions and only needs to learn to extract attributes. Its underperformance compared to `EMMA-LWM` can be attributed to (i) the imperfection of ChatGPT in identifying identities in descriptions (its accuracy is around 90%; see Appendix B) and (ii) the fact that `EMMA-LWM` jointly learns to extract both identity and attribute words, which may be more effective than learning to extract only attribute words.

Evaluation with imaginary trajectories. In this evaluation, for each world model and test trajectory, we reset the model to the initial state of the trajectory and sequentially feed the actions in the trajectory to the model until it predicts the end of the episode. This process generates an imaginary trajectory. We refer to the evaluation trajectory as the real trajectory. We compute precisions of predicting non-zero rewards ($r \neq 0$) and terminations ($d = 1$). To evaluate movement prediction, we compare the distances from the player to an entity in the real and imaginary trajectories. Concretely, let $\delta_{c,t}^{\text{real}}$ and $\delta_{c,t}^{\text{imag}}$ be the Hamming distances from the player to entity c at the t-th time step in a real trajectory τ_{real} and an imaginary trajectory τ_{imag}, respectively. We cal-

Table 2: Results on imaginary trajectory generation. Δ_{dist} measures the similarity between the distances from the player to an entity in a real trajectory and the corresponding imaginary trajectory. The bold number in each column represents the best non-oracle result. `EMMA-LWM` outperforms all baselines in all metrics.

World model	$\Delta_{\text{dist}}(\downarrow)$ NewCombo (easy)	NewAttr (medium)	NewAll (hard)	Non-zero reward precision ($\uparrow$) NewCombo (easy)	NewAttr (medium)	NewAll (hard)	Termination precision ($\uparrow$) NewCombo (easy)	NewAttr (medium)	NewAll (hard)
`Observational`	2.04	2.91	3.00	0.39	0.20	0.15	0.51	0.33	0.28
`Standard`	0.82	1.48	1.68	0.68	0.43	0.50	0.75	0.55	0.62
`GPTHard`	0.89	2.74	2.89	0.75	0.34	0.25	0.79	0.45	0.45
`EMMA-LWM`	**0.57**	**1.14**	**1.29**	**0.88**	**0.69**	**0.70**	**0.88**	**0.75**	**0.71**
`OracleParse`	0.49	0.77	0.92	0.93	0.81	0.77	0.89	0.84	0.79

culate the average difference in a specific time step: $\Delta_{\text{dist}} = \frac{1}{|\mathcal{D}_{\text{eval}}|}\sum_{\tau_{\text{real}}\in\mathcal{D}_{\text{eval}}}\frac{1}{T_{\min}}\sum_{t=1}^{T_{\min}}|\delta_{c,t}^{\text{real}} - \delta_{c,t}^{\text{imag}}|$ where $\mathcal{D}_{\text{eval}}$ is an evaluation split, $T_{\min} = \min(|\tau_{\text{real}}|, |\tau_{\text{imag}}|)$, and τ_{imag} is generated from τ_{real} . For example, for a chasing entity, $\delta_{c,t}^{\text{real}}$ decreases as t increases. If a model mistakenly predicts the entity to be immobile, $\delta_{c,t}^{\text{imag}}$ remains a constant as t progresses. In this case, Δ_{dist} is non-negligible, indicating an error. All evaluation metrics are given in Table 2. The ordering of the models is similar to that in the evaluation with ground-truth trajectories. `EMMA-LWM` is still superior to all baselines in all metrics.

5.3 Application: agents that discuss plans with humans

In this section, we showcase the practicality of our LWM by illustrating that it can facilitate *plan discussions* between an agent and a human supervisor. This approach has the potential to improve the transparency, safety, and performance of real-world agents.

We imagine an agent ordered to perform a task in a previously unseen environment (Figure 1a). Letting the agent perform the task immediately would be extremely risky because of its imperfect knowledge of the environment. Implementing a world model enables the agent to imagine a solution trajectory and present it to a human as a *plan* for review. Conveying plans as trajectories helps the human envision the future behavior of the agent in the real world. Furthermore, the human can improve this behavior by providing feedback to enhance the policy that produces the plan.

A human can update the policy by telling the agent which actions it should have taken. This type of feedback can be incorporated using some form of imitation learning. An agent equipped with a LWM additionally enables the human to **update its policy by giving language feedback that aims to modify its world model**. Although an observational world model also allows this form of adaptation, it requires much more effort from the human to generate the feedback. Concretely, the human has to generate observations in the same format as those in the agent's plan (e.g., they have to draw grids in this setting). Furthermore, many abstract concepts may not be efficiently or precisely specified through non-verbal communication.

We simulate this scenario by placing agents with randomly initialized policies in test environments. These agents are forbidden to interact with the environments. However, they are equipped with world models, which allows for imaginary policy update. The world models are the ones we evaluated in the previous section. Importantly, the models were not trained on any data collected in the environments, simulating the fact that these environments are completely new to the agents.

We train all policies with imitation learning, considering two types of feedback: in *online imitation learning* (Ross et al., 2011), the expert suggests the best actions to take in the states present in the plan; in the *filtered behavior cloning* setting, the expert simply overwrites the agent's plan with their own plan. In the latter setting, the agent chooses the plans that achieve the highest returns according to their world models to imitate. We experiment with a near-optimal expert and a suboptimal expert. We provide more details in Appendix D.

The agents endowed with LWMs can also process language feedback aiming to change their world models. This feedback is simulated by the game manuals accompanying the environments. It serves as the input ℓ of the LWMs. We suppose that a human gives this feedback once to an agent, before adapting it via imitation learning.

We present the performance of the agents after adaptation in Table 3. Learning with the `Observational` world model amounts to the case where the human provides only imitation-learning feedback and cannot adapt the world model via language. Meanwhile, learning with EMMA-LWM represents the case where the human can use language feedback to improve the world model. In all evaluation settings, we observe significant improvements in the average return of policies that adopt our EMMA-LWM. There are still considerable gaps compared to using the OracleParse model, indicating that our model still has room for improvement.

Table 3: Average returns (↑) in real environments of policies trained with imaginary imitation learning using world models. Bold numbers indicate the best non-oracle means in the corresponding settings. An expanded table with all models and details on how the metric was computed are available in Appendix E.

Setting	World model	NewCombo (easy)	NewAttr (medium)	NewAll (hard)
Online IL *(near-optimal)*	Observational	0.75 ± 0.16	-0.41 ± 0.21	-0.21 ± 0.21
	EMMA-LWM (ours)	**1.01** ± 0.12	**0.96** ± 0.17	**0.62** ± 0.21
	OracleParse	1.04 ± 0.13	0.85 ± 0.20	0.91 ± 0.18
Filtered BC *(near-optimal)*	Observational	0.77 ± 0.14	-0.42 ± 0.15	-0.30 ± 0.16
	EMMA-LWM (ours)	**1.18** ± 0.10	**0.75** ± 0.20	**0.44** ± 0.18
	OracleParse	1.17 ± 0.11	0.84 ± 0.19	0.80 ± 0.18
Filtered BC *(suboptimal)*	Observational	0.71 ± 0.15	-0.35 ± 0.18	-0.33 ± 0.17
	EMMA-LWM (ours)	**0.98** ± 0.13	**0.29** ± 0.25	**0.13** ± 0.19
	OracleParse	1.09 ± 0.13	0.50 ± 0.24	0.49 ± 0.18

6 Related work

World models. World models have a rich history dating back to the 1980s (Werbos, 1987). The base architecture has evolved from feed-forward neural networks (Werbos, 1987), to recurrent neural networks (Schmidhuber, 1990a,b, 1991), and most recently, Transformers (Robine et al., 2023; Micheli et al., 2023). In RL settings, world models are the key component of model-based approaches, which train policies in simulation to reduce the amount of interactions with real environments. Model-based RL has been successful in a variety of robotic tasks (Finn and Levine, 2017) and video games (Hafner et al., 2019, 2020, 2023). However, the incorporation of language information into world models has been underexplored. Cowen-Rivers and Naradowsky (2020) propose language-conditioned world models but focus on emergent language rather than human language. Poudel et al. (2023) incorporate features language into the representations of the model. These approaches, however, do not use language to control a world model.

Language-based adaptation. Language information has been incorporated into various aspects of learning. In instruction following (Bisk et al., 2016; Misra et al., 2018; Anderson et al., 2018; Nguyen and Daumé III, 2019), agents are given descriptions of the desired behaviors and learn to interpret them to perform tasks. Language-based learning (Nguyen et al., 2021; Scheurer et al., 2023) employs language-based feedback to train models. Another line of work uses language descriptions of environment dynamics to improve policy learning (Narasimhan et al., 2018; Branavan, 2012; Hanjie et al., 2021; Wu et al., 2023a; Nottingham et al., 2022; Zhong et al., 2020). Rather than using texts to directly improve a policy, our work leverages them to enhance a model of an environment. Recently, several papers propose agents that can read text manuals to play games (Wu et al., 2023a,b). Our work differs from these papers in that we aim to build models that capture exactly the transition function of an environment.

Compositional generalization for language-guided world models. Lin et al. (2024) model a variety of text-augmented environments but do not demonstrate the generalizability of their approach in MESSENGER. Recent work (Zhao et al., 2022; Du et al., 2024; Zhou et al., 2024; Zhang et al., 2024) has developed LWMs with compositional generalizability. While these papers operate on more visually realistic domains than ours, the language they study is simpler, focusing on concepts that correspond to straightforward mappings from input to output such as colors and objects. In contrast, the concepts in MESSENGER are more intricate, regarding interactions among multiple entities.

7 Conclusion

We introduce *Language-Guided World Models*, which can be adapted through natural language. We outline numerous advantages of these models over traditional observational world models. Our model is still lacking in performance and the grid-world environments we experiment with severely underrepresent the real world. Nevertheless, we hope that this work helps envision the potential of LWMs in enhancing the controllability of artificial agents and inspires future efforts to address the compositional generalization challenge.

Acknowledgements

We thank Ameet Deshpande, Vishvak Murahari, and Howard Chen from the Princeton NLP group for valuable feedback, comments, and discussions. We thank Kurtland Chua for helpful feedback. This material is based upon work supported by the National Science Foundation under Grant No. 2239363. Any opinions, findings, and conclusions or recommendations expressed in this material are those of the author(s) and do not necessarily reflect the views of the National Science Foundation.

References

Peter Anderson, Qi Wu, Damien Teney, Jake Bruce, Mark Johnson, Niko Sünderhauf, Ian Reid, Stephen Gould, and Anton Van Den Hengel. 2018. Vision-and-language navigation: Interpreting visually-grounded navigation instructions in real environments. In *Proceedings of the IEEE conference on computer vision and pattern recognition*, pages 3674–3683.

Yonatan Bisk, Deniz Yuret, and Daniel Marcu. 2016. Natural language communication with robots. In *Proceedings of the 2016 Conference of the North American Chapter of the Association for Computational Linguistics: Human Language Technologies*, pages 751–761.

SRK Branavan. 2012. Learning to win by reading manuals in a monte-carlo framework. *Journal of Artificial Intelligence Research*, 43:661–704.

Rahma Chaabouni, Roberto Dessì, and Eugene Kharitonov. 2021. Can transformers jump around right in natural language? assessing performance transfer from scan. In *BlackboxNLP workshop (EMNLP)*.

Kurtland Chua, Roberto Calandra, Rowan McAllister, and Sergey Levine. 2018. Deep reinforcement learning in a handful of trials using probabilistic dynamics models. *Advances in neural information processing systems*, 31.

Alexander I Cowen-Rivers and Jason Naradowsky. 2020. Emergent communication with world models. *arXiv e-prints*, pages arXiv–2002.

David Dalrymple, Joar Skalse, Yoshua Bengio, Stuart Russell, Max Tegmark, Sanjit Seshia, Steve Omohundro, Christian Szegedy, Ben Goldhaber, Nora Ammann, et al. 2024. Towards guaranteed safe ai: A framework for ensuring robust and reliable ai systems. *arXiv preprint arXiv:2405.06624*.

Marc Deisenroth and Carl E Rasmussen. 2011. Pilco: A model-based and data-efficient approach to policy search. In *Proceedings of the 28th International Conference on machine learning (ICML-11)*, pages 465–472.

Yilun Du, Sherry Yang, Bo Dai, Hanjun Dai, Ofir Nachum, Josh Tenenbaum, Dale Schuurmans, and Pieter Abbeel. 2024. Learning universal policies via text-guided video generation. *Advances in Neural Information Processing Systems*, 36.

Nouha Dziri, Ximing Lu, Melanie Sclar, Xiang Lorraine Li, Liwei Jian, Bill Yuchen Lin, Peter West, Chandra Bhagavatula, Ronan Le Bras, Jena D Hwang, et al. 2023. Faith and fate: Limits of transformers on compositionality. In *Proceedings of Advances in Neural Information Processing Systems*.

Chelsea Finn and Sergey Levine. 2017. Deep visual foresight for planning robot motion. In *2017 IEEE International Conference on Robotics and Automation (ICRA)*, pages 2786–2793. IEEE.

David Ha and Jürgen Schmidhuber. 2018. World models. *arXiv preprint arXiv:1803.10122*.

Danijar Hafner, Timothy Lillicrap, Jimmy Ba, and Mohammad Norouzi. 2019. Dream to control: Learning behaviors by latent imagination. *arXiv preprint arXiv:1912.01603*.

Danijar Hafner, Timothy Lillicrap, Mohammad Norouzi, and Jimmy Ba. 2020. Mastering atari with discrete world models. *arXiv preprint arXiv:2010.02193*.

Danijar Hafner, Jurgis Pasukonis, Jimmy Ba, and Timothy Lillicrap. 2023. Mastering diverse domains through world models. *arXiv preprint arXiv:2301.04104*.

Austin W Hanjie, Victor Y Zhong, and Karthik Narasimhan. 2021. Grounding language to entities and dynamics for generalization in reinforcement learning. In *International Conference on Machine Learning*, pages 4051–4062. PMLR.

Yichen Jiang and Mohit Bansal. 2021. Inducing transformer's compositional generalization ability via auxiliary sequence prediction tasks. In *Proceedings of Empirical Methods in Natural Language Processing*.

Daniel Keysers, Nathanael Schärli, Nathan Scales, Hylke Buisman, Daniel Furrer, Sergii Kashubin, Nikola Momchev, Danila Sinopalnikov, Lukasz Stafiniak, Tibor Tihon, et al. 2020. Measuring compositional generalization: A comprehensive method on realistic data. In *Proceedings of the International Conference on Learning Representations*.

Jessy Lin, Yuqing Du, Olivia Watkins, Danijar Hafner, Pieter Abbeel, Dan Klein, and Anca Dragan. 2024. Learning to model the world with language. In *Proceedings of the International Conference of Machine Learning*.

Ilya Loshchilov and Frank Hutter. 2017. Decoupled weight decay regularization. *arXiv preprint arXiv:1711.05101*.

Vincent Micheli, Eloi Alonso, and François Fleuret. 2023. Transformers are sample-efficient world models. In *Proceedings of the International Conference on Learning Representations.*

Dipendra Misra, Andrew Bennett, Valts Blukis, Eyvind Niklasson, Max Shatkhin, and Yoav Artzi. 2018. Mapping instructions to actions in 3d environments with visual goal prediction. *arXiv preprint arXiv:1809.00786.*

Karthik Narasimhan, Regina Barzilay, and Tommi Jaakkola. 2018. Grounding language for transfer in deep reinforcement learning. *Journal of Artificial Intelligence Research*, 63:849–874.

Khanh Nguyen and Hal Daumé III. 2019. Help, anna! visual navigation with natural multimodal assistance via retrospective curiosity-encouraging imitation learning. *arXiv preprint arXiv:1909.01871.*

Khanh X Nguyen, Dipendra Misra, Robert Schapire, Miroslav Dudík, and Patrick Shafto. 2021. Interactive learning from activity description. In *International Conference on Machine Learning*, pages 8096–8108. PMLR.

Kolby Nottingham, Alekhya Pyla, Sameer Singh, and Roy Fox. 2022. Learning to query internet text for informing reinforcement learning agents. *arXiv preprint arXiv:2205.13079.*

Long Ouyang, Jeffrey Wu, Xu Jiang, Diogo Almeida, Carroll Wainwright, Pamela Mishkin, Chong Zhang, Sandhini Agarwal, Katarina Slama, Alex Ray, et al. 2022. Training language models to follow instructions with human feedback. *Advances in Neural Information Processing Systems*, 35:27730–27744.

Rudra PK Poudel, Harit Pandya, Chao Zhang, and Roberto Cipolla. 2023. Langwm: Language grounded world model. *arXiv preprint arXiv:2311.17593.*

Jan Robine, Marc Höftmann, Tobias Uelwer, and Stefan Harmeling. 2023. Transformer-based world models are happy with 100k interactions. In *Proceedings of the International Conference on Learning Representations.*

Stéphane Ross, Geoffrey Gordon, and Drew Bagnell. 2011. A reduction of imitation learning and structured prediction to no-regret online learning. In *Proceedings of the fourteenth international conference on artificial intelligence and statistics*, pages 627–635. JMLR Workshop and Conference Proceedings.

Jérémy Scheurer, Jon Ander Campos, Tomasz Korbak, Jun Shern Chan, Angelica Chen, Kyunghyun Cho, and Ethan Perez. 2023. Training language models with language feedback at scale. *arXiv preprint arXiv:2303.16755.*

Jürgen Schmidhuber. 1990a. *Making the world differentiable: on using self supervised fully recurrent neural networks for dynamic reinforcement learning and planning in non-stationary environments*, volume 126. Inst. für Informatik.

Jürgen Schmidhuber. 1990b. An on-line algorithm for dynamic reinforcement learning and planning in reactive environments. In *1990 IJCNN international joint conference on neural networks*, pages 253–258. IEEE.

Jürgen Schmidhuber. 1991. A possibility for implementing curiosity and boredom in model-building neural controllers. In *Proc. of the international conference on simulation of adaptive behavior: From animals to animats*, pages 222–227.

Jürgen Schmidhuber. 2015. On learning to think: Algorithmic information theory for novel combinations of reinforcement learning controllers and recurrent neural world models. *arXiv preprint arXiv:1511.09249.*

Theodore R Sumers, Mark K Ho, Robert D Hawkins, and Thomas L Griffiths. 2023. Show or tell? exploring when (and why) teaching with language outperforms demonstration. *Cognition*, 232:105326.

Ashish Vaswani, Noam Shazeer, Niki Parmar, Jakob Uszkoreit, Llion Jones, Aidan N Gomez, Łukasz Kaiser, and Illia Polosukhin. 2017. Attention is all you need. *Advances in neural information processing systems*, 30.

Paul J Werbos. 1987. Learning how the world works: Specifications for predictive networks in robots and brains. In *Proceedings of IEEE International Conference on Systems, Man and Cybernetics, NY.*

Yue Wu, Yewen Fan, Paul Pu Liang, Amos Azaria, Yuanzhi Li, and Tom Mitchell. 2023a. Read and reap the rewards: Learning to play atari with the help of instruction manuals. In *Workshop on Reincarnating Reinforcement Learning at ICLR 2023.*

Yue Wu, So Yeon Min, Shrimai Prabhumoye, Yonatan Bisk, Ruslan Salakhutdinov, Amos Azaria, Tom Mitchell, and Yuanzhi Li. 2023b. Spring: Studying papers and reasoning to play games. In *Thirty-seventh Conference on Neural Information Processing Systems.*

Hongxin Zhang, Zeyuan Wang, Qiushi Lyu, Zheyuan Zhang, Sunli Chen, Tianmin Shu, Yilun Du, and Chuang Gan. 2024. Combo: Compositional world models for embodied multi-agent cooperation. *arXiv preprint arXiv:2404.10775.*

Linfeng Zhao, Lingzhi Kong, Robin Walters, and Lawson LS Wong. 2022. Toward compositional generalization in object-oriented world modeling. In *International Conference on Machine Learning*, pages 26841–26864. PMLR.

Ruijie Zheng, Khanh Nguyen, Hal Daumé III, Furong Huang, and Karthik Narasimhan. 2023. Progressively efficient learning. *arXiv preprint arXiv:2310.13004.*

Victor Zhong, Austin W. Hanjie, Sida I. Wang, Karthik Narasimhan, and Luke Zettlemoyer. 2021. Silg: The multi-environment symbolic interactive language grounding benchmark. In *Neural Information Processing Systems (NeurIPS)*.

Victor Zhong, Tim Rocktäschel, and Edward Grefenstette. 2020. Rtfm: Generalising to new environment dynamics via reading. In *International Conference on Learning Representations*.

Siyuan Zhou, Yilun Du, Jiaben Chen, Yandong Li, Dit-Yan Yeung, and Chuang Gan. 2024. Robodreamer: Learning compositional world models for robot imagination. *arXiv preprint arXiv:2404.12377*.

A GPTHard model

This approach leverages the language-understanding capabilities of ChatGPT. Through few-shot prompting, we instruct this model to determine the identity of the entity mentioned in each manual description. In this approach, we generate only the set of values $\boldsymbol{m}^{\text{val}}$ as in Eq 1. Instead of learning soft attention, we directly route the values to the identity embeddings. Concretely, the feature vector added to $\boldsymbol{i}_t^c$ in Eq 3 is $\boldsymbol{z}_t^c = \boldsymbol{m}_{j_c}^{\text{val}}$ where j_c is the index of the description that mentions entity c according to ChatGPT.

We compose the following prompt for parsing descriptions. We use the "May 3, 2023" release of ChatGPT. We feed to the model one description at a time instead of a whole manual of three descriptions. We ask it to also extract the role and movement pattern, but use only the parsed identity in the `GPTHard` model. The "ChatGPT identity-parsing" column in Table 4 shows the fraction of games in each split in which ChatGPT correctly identifies all three identities in a game. Note that the `OracleParse` model uses the ground-truth parses rather than these parses.

````
You are playing a role-playing video
    game where you will need to read
    textual descriptions to figure out
    the attributes of a character.

This is a list of characters and their
    corresponding IDs:
airplane: 2
mage: 3
dog: 4
bird: 5
fish: 6
scientist: 7
thief: 8
ship: 9
ball: 10
robot: 11
queen: 12
sword: 13

This is a list of movement types and
    their corresponding IDs:
chasing: 0
fleeing: 1
stationary: 2

This is a list of role types and their
    corresponding IDs:
dangerous enemy: 0
secret message: 1
essential objective: 2

Now, read a description and tell me
    which character is being mentioned
    and what are its movement type and
    role type. Your answer should follow
     this format:
```
Answer: Character ID, movement type ID,
    role type ID
```

Here are a few examples:

Description: the plane that's flying
    near where you are is the critical
    objective.
Answer: 2, 0, 2

Description: the escaping humanoid is an
     important goal.
Answer: 11, 1, 2

Description: the mage is inching near
    you is a lethal opponent.
Answer: 3, 0, 0

Description: the classified document is
    the hound coming your way.
Answer: 4, 0, 1

Description: the important goal is the
    orb which is creeping close to you.
Answer: 10, 0, 2

Now provide the answer for the following
     description. Follow the format of
    the previous answers:

Description: [PLACEHOLDER]
````

B Dataset

Statistics of our dataset are provided in Table 4. The maximum trajectory length is 32. We implement five rule-based behavior policies: `survive` (avoid the enemy and goal), `win the game`, `suicide` (go to the enemy), `obtain the message`, and `act randomly`. The `survive` policy acts randomly when the distances to the enemy and the goal are greater than or equal to 6. Otherwise, it takes the action that makes its distance to those entities at least 3. If that is impossible, it chooses the action that maximizes the minimum distance to one of the two entities. The `win the game` policy is not optimal: it simply aims to obtain the message and then run to the goal, without having a strategy to avoid the enemy. We run a breadth-first search to find the next best action to get to an entity.

For the training split, we generate 66 trajectories per game. The behavior policy for each trajectory is chosen uniformly randomly among the five rule-based policies. For each evaluation split, we generate 5 trajectories per game, using every rule-based policy to generate trajectories.

Split		Unique games	Unique descriptions	Trajectories	ChatGPT identity-parsing accuracy (%)
Train		1,536	986	101,376	92
Dev	NewCombo	896	598	4,480	89
	NewAttr	204	319	1,020	88
	NewAll	856	1,028	4,280	86
Test	NewCombo	896	587	4,480	90
	NewAttr	204	306	1,020	93
	NewAll	856	1,016	4,280	88

Table 4: MESSENGER data statistics. The last column shows the fraction of games in each split in which ChatGPT correctly identifies all three identities in a game.

Hyperparameter	Value
Hidden size	256
Number of encoder layers	4
Number of decoder layers	4
Number of decoder token blocks	33
Dropout rate	0.1
Batch size	32
Number of training batches	100K
Evaluation every	500 batches
Optimizer	AdamW
Learning rate	1e-4
Max. gradient norm	10

Table 5: Training hyperparameters.

C Training details

Our implementation of Transformer is largely based on the IRIS codebase (Micheli et al., 2023).[2] We implement cross-attention for the `Standard` baseline, and EMMA for our model.

Initialization. We find that the default PyTorch initialization scheme does not suffice for our model to generalize compositionally. We adopt the following initialization scheme from the IRIS codebase:

```
def init_weights(module):
    if isinstance(module, (nn.Linear, nn.Embedding)):
        module.weight.data.normal_(mean=0.0, std=0.02)
        if isinstance(module, nn.Linear) and module.bias is not None:
            module.bias.data.zero_()
    elif isinstance(module, nn.LayerNorm):
        module.bias.data.zero_()
        module.weight.data.fill_(1.0)
```

which is evoked by calling `self.apply(init_weights)` in the model's constructor. We initialize all models with this scheme, but only `EMMA-LWM` and `OracleParse` perform well consistently on various random seeds.

Compute resources. Experiments were primarily run on a cluster of NVIDIA RTX2080 GPUs, and each experiment was run on a single device. To generate Table 1, we trained each world model for 24 GPU hours, 5 seeds each. To generate Table 3 and 6, we trained each of the 5 world models on each of the 90 games (3 difficulties for 30 game configurations) using the 3 different downstream policy training strategies, with each game being 12 GPU hours.

D Imitation learning experiments

The learning policy follows the EMMA-based policy architecture of (Hanjie et al., 2021), which at each time step processes a stack of 3 most recent observations with a convolution-then-MLP encoder. We train the policy with 2,000 batches using the same optimizer hyperparameters as those of the world models.

For the online IL setting, we use the `win the game` rule-based policy (Appendix B) as the expert. For the filtered BC setting, we train an EMMA policy to overfit the test environment. We then use a fully converged checkpoint of the policy as the near-optimal expert, and a not fully converged checkpoint as the suboptimal expert. The former is trained for 10,000 iterations and the latter is trained for 2,000 iterations.

The test environments are randomly chosen from the test splits. We select 10 environments per split. We evaluate each policy for 48 episodes in the real environment. These episodes cover all 24 initial configurations of a stage-two MESSENGER game.

E Extended results

Figure 4 studies the performance of the models when conditioned on prefixes of the ground-truth

[2]https://github.com/eloialonso/iris

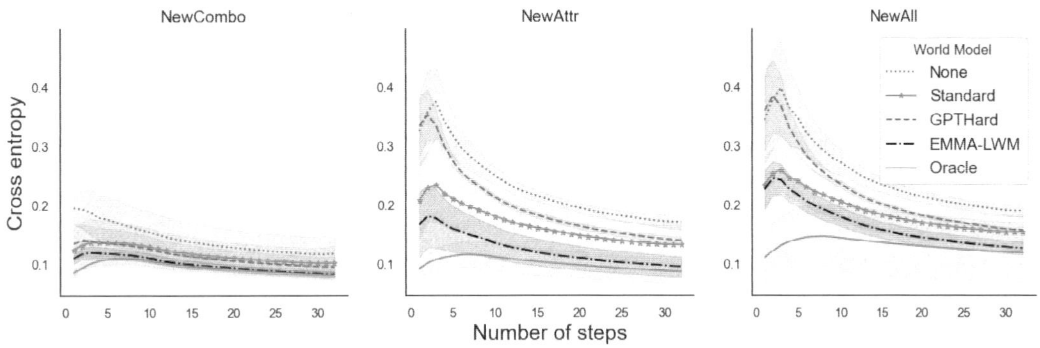

Figure 4: The cross entropy losses of the models when conditioned on ground-truth trajectory prefixes up to a certain length. We plot the means with 95% t-value confidence intervals. The losses generally decrease as the prefix length increases. EMMA-LWM outperforms baselines given any prefix length.

trajectories. The losses of all models decrease as the prefix length increases, but the baselines cannot close the gaps with EMMA-LWM. Across all splits, EMMA-LWM conditioned on a one-step history outperforms Observational conditioned on one third of a ground-truth trajectory, demonstrating that our model has effectively leveraged the textual information.

Table 6 presents the results of all the models in the simulation of plan discussion (§5.3).

Table 6: Average returns (↑) in real environments of policies trained with imaginary imitation learning using world models. For each world model type, we use the best checkpoint of a run chosen randomly among the five runs mentioned in Table 1. Experiments are conducted in 90 environments randomly chosen from the test splits (30 from each split). For each environment and learned policy, we compute the average return over 48 runs. For each split, we report the means of the average returns in the 30 environments with 95% t-value confidence intervals. Bold numbers indicate the best non-oracle means in the corresponding settings. EMMA-LWM outperforms all baselines in all settings.

Setting	World model	NewCombo (easy)	NewAttr (medium)	NewAll (hard)
Online IL *(near-optimal expert)*	Observational	0.75 ± 0.16	-0.41 ± 0.21	-0.21 ± 0.21
	Standard	0.93 ± 0.13	0.04 ± 0.26	0.30 ± 0.22
	GPTHard	0.82 ± 0.15	-0.20 ± 0.20	-0.06 ± 0.21
	EMMA-LWM (ours)	**1.01 ± 0.12**	**0.96 ± 0.17**	**0.62 ± 0.21**
	OracleParse	1.04 ± 0.13	0.85 ± 0.20	0.91 ± 0.18
Filtered BC *(near-optimal expert)*	Observational	0.77 ± 0.14	-0.42 ± 0.15	-0.30 ± 0.16
	Standard	1.05 ± 0.14	0.20 ± 0.27	0.17 ± 0.20
	GPTHard	0.79 ± 0.15	-0.10 ± 0.20	-0.07 ± 0.20
	EMMA-LWM (ours)	**1.18 ± 0.10**	**0.75 ± 0.20**	**0.44 ± 0.18**
	OracleParse	1.17 ± 0.11	0.84 ± 0.19	0.80 ± 0.18
Filtered BC *(suboptimal expert)*	Observational	0.71 ± 0.15	-0.35 ± 0.18	-0.33 ± 0.17
	Standard	0.68 ± 0.15	-0.15 ± 0.21	-0.10 ± 0.17
	GPTHard	0.75 ± 0.22	0.05 ± 0.25	0.06 ± 0.17
	EMMA-LWM (ours)	**0.98 ± 0.13**	**0.29 ± 0.25**	**0.13 ± 0.19**
	OracleParse	1.09 ± 0.13	0.50 ± 0.24	0.49 ± 0.18

Learning Communication Policies for Different Follower Behaviors in a Collaborative Reference Game

Philipp Sadler[1], Sherzod Hakimov[1], David Schlangen[1,2]

[1]CoLabPotsdam / Computational Linguistics
Department of Linguistics, University of Potsdam, Germany
[2]German Research Center for Artificial Intelligence (DFKI), Berlin, Germany
Correspondence: firstname.lastname@uni-potsdam.de

Abstract

In this work, we evaluate the adaptability of neural agents towards assumed partner behaviors in a collaborative reference game. In this game, success is achieved when a knowledgeable guide can verbally lead a follower to the selection of a specific puzzle piece among several distractors. We frame this language grounding and coordination task as a reinforcement learning problem and measure to which extent a common reinforcement training algorithm (PPO) is able to produce neural agents (the guides) that perform well with various heuristic follower behaviors that vary along the dimensions of confidence and autonomy. We experiment with a learning signal that in addition to the goal condition also respects an assumed communicative effort. Our results indicate that this novel ingredient leads to communicative strategies that are less verbose (staying silent in some of the steps) and that with respect to that the guide's strategies indeed adapt to the partner's level of confidence and autonomy.

Figure 1: An exemplary interaction between a guide and a follower who controls the gripper (the black dot). The guide observes the scene v_0 and refers to a piece initially with l_0. The follower has only a partial view p_0 (the grey box) and might go wrong. The guide can provide further information based on the follower's actions until a piece is selected at time step T. The guide should learn that fewer utterances are necessary with a more autonomous and confident follower.

1 Introduction

Sometimes we feel like we could continue another person's sentence. This happens in particular with people we know well or we often interact with. A common phrase coined to this phenomenon is that "people are on the same wavelength". Indeed Davidesco et al. (2023) found that brain activities somewhat synchronize between teachers and students during lessons. Even more surprising, synchronicity becomes a good predictor of the learning success of the students. A psycho-linguistic study by Clark and Wilkes-Gibbs (1986) observed the language use of collaborative partners during an ongoing goal-oriented interaction: They (implicitly) agree on newly introduced noun phrases and a common strategy to achieve the goal together. Interestingly, the number of used words drastically decreases during the collaboration. The participants strive towards reduced individual efforts while the number of successful outcomes stays high. We see that human-human interaction is characterized by synchronicity (adaption) and the reduction of individual efforts. Still, the modelling of changing behaviors (or different others) remains an open problem "due to the essentially unconstrained nature of what other agents may do" (Albrecht and Stone, 2018). Are neural agents capable of adapting to their interactants and converge to useful strategies when the partner's behavior becomes apparent only during an ongoing interaction itself?

In this work, we frame a collaborative language coordination and grounding task (see Figure 1) as a reinforcement learning problem (Sutton and

Proceedings of the 4th Workshop on Spatial Language Understanding and Grounded Communication for Robotics (SpLU-RoboNLP 2024), pages 17–29
August 11-16, 2024 ©2024 Association for Computational Linguistics

Barto, 2018) and evaluate, if and to which extent a common training algorithm Proximal Policy Optimization (PPO) (Schulman et al., 2017) is able to produce neural agents that perform well with a variety of partner behaviors. To study how learning agents potentially adapt to an assumed partner's behavior, we propose a challenging vision and language grounding task where two players have to coordinate on the selection of a puzzle piece (a Pentomino, a shape of five adjacent squares; Golomb (1996)) among several distractors while (i) the actual target piece is only known to one of them (the guide), and (ii) only the other can perform the selection (the follower).

The main idea is that we assume an ongoing interaction in which the follower's behavior changes. After some time the follower should become more autonomous and more confident in choosing actions and executing its own plan (as pointed out by Clark and Wilkes-Gibbs (1986)). But instead of treating this as a multi-agent setting directly, we follow Yang et al. (2022) with the notion of assigning different agents to different sub-tasks and learn a policy for each of the controllable follower behaviors (the sub-tasks) separately. The resulting policies represent a guide's communicative strategy at certain points in time of the assumed ongoing interaction.

Our expectations on the learned communicative strategies of the guide are that in the beginning (with a less autonomous, less confident follower) more is to be said. Later on, with a more autonomous and confident follower, the guide learns that it "does not need to say anything" to be successful and consequently reduces its effort. Our contributions are as follows[1]:

- We propose a challenging RL environment: a reference game in which a neural agent (the guide) has to learn communication strategies that are **successful and reduce an assumed effort**, and
- contribute a plausible follower policy (the training partner) that is variable on two dimensions: **confidence** and **autonomy**, and
- present strong baseline guide policies for this difficult cooperative reference game that are indeed able to balance out episode success and their individual effort by **learning to stay silent**.

[1]Source code is publicly available at: https://github.com/clp-research/different-follower-behaviors

2 Related Work

Vision and language navigation. The use of natural language to guide an instruction following agent has been heavily studied for the vision and language navigation task (Gu et al., 2022; Nguyen et al., 2019; Nguyen and Daumé III, 2019; Fried et al., 2018; Thomason et al., 2019). For example, Nguyen and Daumé III (2019) train an instruction giver (IG) on a pre-collected dataset of instructions. The follower is then allowed to ask the IG for more information during task execution. Although the setting is very similar, but in our work the guide has to learn when to provide more information to the follower. In our setting, the language back-channel for the follower is cut, so the players must use the vision signal in their coordination and the guide's must monitor the follower's behavior.

Natural language goals in RL. Using natural language to describe the goal state in an RL problem has become a common theme (Chevalier-Boisvert et al., 2019; Gao et al., 2022; Padmakumar et al., 2022; Pashevich et al., 2021; Suhr and Artzi, 2023). This research direction is interesting because it could allow humans to interact more easily with learned agents. There is work that shows that intermediate language inputs are a valuable signal in task-oriented visual environments (Co-Reyes et al., 2019; Mu et al., 2022). Indeed Huang et al. (2023) found that natural language can "provide a gradient" towards the goal state. But they also point out the "brittleness" of these signals because the language input might align badly with sub-trajectories. A key challenge here is the variability of expressions in language that can be produced and understood in the defined action space. Even in relatively simple environments, there might arise an overwhelming amount of situations for an agent to handle (Chevalier-Boisvert et al., 2019). We weaken the action space exploration problem by using ideas from natural language understanding (Moon et al., 2020; E et al., 2019) and let the guide produce language actions in a well-defined reduced "intent space". These intents are then verbalized (using templates; which could be a conditioned pre-trained language model) and given to the follower.

Interactive sub-goal generation in RL. Sun et al. (2023) use a pre-trained large language model to generate possible plans (in the form of source code) for the completion of a task. The learning process is extended with a mechanism that allows

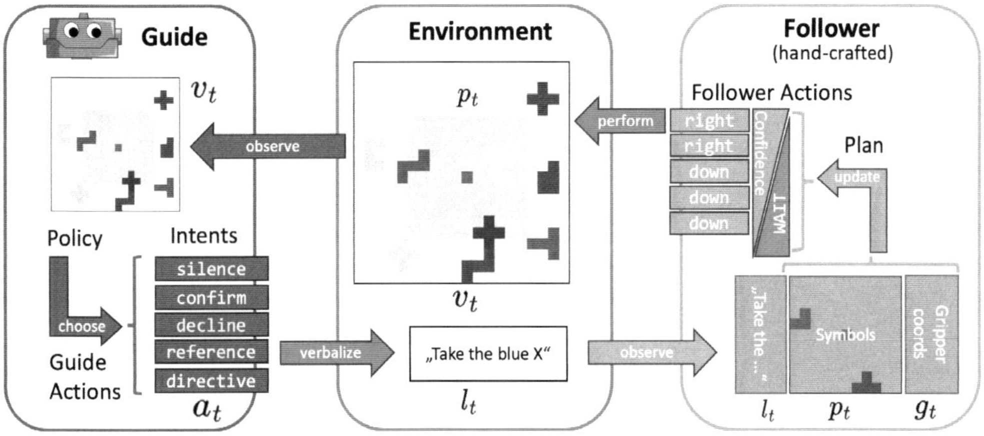

Figure 2: The general information and decision-making flow of the reference game. The guide observes v_t which contains the full scene in pixel space and additionally the gripper position (4th-channel) and target piece (5th-channel). Given this, the guide chooses an intent action a_t that gets verbalized into a template-based sentence l_t. Then, the follower receives the utterance l_t, the gripper coordinate g_t and a symbolic representation of a partial view of the scene p_t. The hand-crafted policy updates the plan accordingly based on its given representation of the world. Finally, the follower's next planned action (or `wait`) is performed with a certain chance defined by the attached confidence. The process repeats until a piece is taken or time runs out.

the model to learn the refinement of single actions or an entire plan respectively. Indeed neural agents perform better when they self-predict sub-goals to be achieved (with an intrinsic reward) instead of reaching for the final goal immediately (Jurgenson and Tamar, 2023; Chane-Sane et al., 2021; Pertsch et al., 2020; Jeon et al., 2022). For example, Lee and Kim (2023) study the task of finding the best route in a simple visual domain by training a sub-goal system that predicts intermediate coordinates. In contrast to them, our guiding agent has to produce utterances to describe a sub-goal (and we use referring expressions or directions). Gürtler et al. (2021) also address the question of "when to provide sub-goals", which is necessary in our task. Nevertheless, in distinction to these works, we treat the sub-goal generation not just as additional information for the follower's success but are interested in the learned communicative strategies themselves. We treat the sub-goal providing guide as an individual participant in the environment similar to a multi-agent setting.

Skill learning in cooperative multi-agent RL. We treat both guide and follower as agents in a cooperative setting and follow work that uses hand-crafted policies (Wang et al., 2021; Ghosh et al., 2020; Xie et al., 2020). In this sense, our approach is similar to heterogeneous skill learning (Chang et al., 2022; Liu et al., 2022; Hu et al., 2023) where a single agent is trained to acquire a variety of skills (in our case communication strategies). This is, in particular, helpful due to the differences in the action spaces of the guide (language acts) and the follower (movements). In addition, this method of having a hand-crafted follower policy allows us to avoid the problem of emergent communication where agents agree on a language that becomes inaccessible to humans (Lowe et al., 2019).

3 The Collaborative Reference Game

We use a collaborative game of referential and interactive language with Pentomino pieces (Sadler et al., 2023) and extend it for guidance learning. A guide has to instruct a follower to select a specific target piece with a gripper. In this setting, both players are constrained as follows: The guide can provide utterances but cannot move the gripper. The follower can move the gripper but is not allowed to provide an utterance. This asymmetry in knowledge and skill forces them to work together and coordinate. Zarrieß et al. (2016) found that such a reference game leads to diverse language use on the guide's side.

3.1 Problem Formulation

We frame this game as an RL problem with sparse rewards. At each time-step t, given an observation $o_t \in \mathcal{O}$ of the environment (see Figure 2), the guide has to choose an action a_t such that the overall resulting sequence of actions $(a_0, ..., a_t, ..., a_T)$

(which become verbalized into $(l_0, ..., l_t, ..., l_T)$) maximizes the sparse reward $\mathcal{R}(o_T) = r$ that is given on episode end, either when a piece is selected by the follower or t reaches $T_{max} = 30$. This maximal number of steps is sufficient to navigate to the target piece with some extra steps for corrections on our 21×21 tile maps. The follower starts in the center of the map so that the farthest tile would be 10 horizontal plus 10 vertical steps away.

3.2 Actions

We let the guide predict "intent" actions and translate them into sentences instead of predicting words directly to reduce the agent's burden on action space exploration (later this verbalization process could be done by a language generation system). Here we focus on the guide's choice among five intent categories: `silence`, `confirm`, `decline`, `directive`, `reference`. For the `directives`, we allow more fine-grained control over the utterance production, so that the agent has to choose between `left`, `right`, `up`, `down` and `take`. Similarly, for the `references` the agent has to choose among possible preference orders `PCS`, `PSC`, `SPC`, `CPS`, `SCP` and `CSP` (in which `P`, `C` and `S` stand for piece, color, and shape, respectively). These preference orders (PO) define the order in which properties are compared between the target piece and its distractors. This means, for example, that a `CSP`-based reference is likely to mention the target piece's color because the color is tried first to distinguish the target from its distractors (and it is very unlikely that all pieces share the same color). These six `reference` actions, five `directive` actions, `silence`, `confirm` and `decline` lead to a total of $|A| = 14$ actions. In comparison, the vocabulary contains 37 tokens and the maximal sentence length is 12 which results in 37^{12} possible utterances when predicting individual words instead of intents.

3.3 Verbalization

The chosen intent is then verbalized based on templates by application of the following rules:

```
silence → <empty string>
confirm → Yes this [way|<piece>]
decline → Not this [way|<piece>]
directive(take) → Take <piece>
directive(dir) → Go <dir>
reference(PO) → Take the <IA(PO)>
```

where `<piece>` resolves to a piece's color and shape when the current gripper position is located over a piece (or otherwise simply `piece`). The direction `<dir>` resolve to the according intent name. The fine-grained reference intent (`PO`) is given to the "Incremental Algorithm" (Dale and Reiter, 1995), which produces the referring expression for reference verbalization (see Appendix A.1).

3.4 Rewards

Following Chevalier-Boisvert et al. (2019), we define a basic sparse reward for playing the game:

$$\mathcal{R}_{\text{Game}} = 1 - 0.9 * (T/T_{\text{max}}) \tag{1}$$

In addition, we introduce a sparse reward for the guide's individual effort in an episode:

$$\mathcal{R}_{\text{Guide}} = 1 - 0.9 * (E_{\text{Guide}}/T_{\text{max}}) \tag{2}$$

where the guide's effort E_{Guide} is the sum over the assumed efforts of taking the respective actions:

$$E_{\text{Guide}} = \sum_{t=1}^{T} \begin{cases} 0, & \text{if } a_t \in \{\texttt{silence}\} \\ 1.0, & \text{if } a_t \in \{\texttt{confirm}, \texttt{decline}\} \\ 1.1, & \text{if } a_t \in \{\texttt{directive}\} \\ 1.2, & \text{if } a_t \in \{\texttt{reference}\} \end{cases} \tag{3}$$

These action-based efforts follow the assumed cognitive load for producing them i.e. saying nothing is the cheapest and comparing pieces with each other to produce a reference is the highest. Finally, we give an additional reward ($\mathcal{R}_{\text{Outcome}}$) of $+1$ when the correct piece or a penalty of -1 if the wrong or no piece has been taken at all, so that:

$$\mathcal{R} = (\mathcal{R}_{\text{Game}} + \mathcal{R}_{\text{Guide}})/2 + \mathcal{R}_{\text{Outcome}} \tag{4}$$

Given this formulation, the guide has to play the game by being active (not just stay silent), achieve the goal (get the bonus) and reduce its individual effort (stay mostly silent) to reach a high reward.

3.5 Observations

The environment exposes at each time-step t an observation o_t that contains the following:

- the follower's gripper coordinates $g_t = (x, y)$
- the guide's utterance l_t (might be empty)
- a full view of the scene v_t for the guide
- a partial view p_t of the scene for the follower

The visual observations are 3-dimensional representations of the full $W \times H$-sized board for the guide (RGB-images) and a 11×11-sized cut-out centered on the gripper's position for the follower (CSI-images). We add a 4th channel to the visual observations to indicate the gripper position by setting the values to zero at g_t and one otherwise. In addition, the guide is informed about the target piece coordinates by setting the according values to zero for the target piece and ones otherwise on a 5th channel of its visual observation. For our purposes, the follower receives a symbolic representation of the partial view where colors, shapes and piece IDs are mapped to numbers (see Appendix A.1).

3.6 Task Instances

The task is that a guide provides utterances to a follower who has to take an intended target piece among several other pieces (the distractors). Thus, a game instance of this task is defined by the number and identity of pieces on the board, including which of these is the target piece, and by the size of the board.

The appearance and positioning of the pieces is derived from symbolic piece representations: a tuple of shape (9), color (6), and position (8). We experiment with 360 of these symbolic pieces which include all shapes, colors, and positions and split them into distinct sets (see Table 1). Therefore, the target symbols for the testing tasks are distinct from the ones seen during training (they might share color and shape though, but are for example positioned elsewhere).

We ensure the reproducibility of our experiments by constructing 2500 training, 175 validation, and 420 testing tasks representing scenes with a map size of 21×21 tiles (see Appendix A.2 for the detailed generation process) where each piece occupies five adjacent tiles and overlapping is avoided.

	TPS	Tasks	Boards
Training	275	2500	700
Validation	25	175	175
Testing	60	420	420

Table 1: The number of tasks and boards in each data split. The target pieces for the tasks are chosen from non-overlapping sub-sets of target piece symbols (TPS). For evaluation splits, we mix-in training pieces as distractors. We construct boards with at least 1 and up to 7 distractors.

4 The Follower Behaviors

For the follower, we take inspiration from Sun et al. (2023) who suggest a plan-based approach towards solving text-based tasks with language models: given a task's natural language instruction their model initially produces a plan, which is then executed and repeatedly refined or revised. We implement a policy that keeps track of a plan that contains up to 10 actions (the plan horizon; which is exactly the number of actions needed to reach the diagonal corner of the partial view). Our follower's behavior of following the plan is adjustable along two dimensions: confidence and autonomy.

Confidence. The actions in the plan are associated with a decreasing probability of being executed (the "confidence triangle" in Figure 2) so that given a discount factor $\phi \in [0, 1]$ and a lower threshold $\mathrm{L} \in [0, 1]$ we calculate:

$$\text{Confidence}(a_i) = \max(\phi^i, \mathrm{L}) \tag{5}$$

Which introduces a notion of confidence: either the planned action is executed or a wait action occurs (hesitation). Furthermore, this conceptualizes that a follower becomes increasingly unsure about the continuation of the plan without receiving feedback from the guide.

Autonomy. The revision process for our follower policy is conceptually divided into five subprograms that run after the guide's utterance is received, parsed and the assumed intent type is determined, as follows:

- `on_silence`: The follower executes, based on confidence, the next action in the plan (if available). Otherwise, it waits.
- `on_confirm`: The follower sets the confidence for all actions in the current plan to 1. Then the next action is chosen as described under `on_silence`.
- `on_decline`: The follower erases the current plan. As the plan is then empty, a wait action will be returned.
- `on_directive`: The follower parses the utterances for the concrete directives (a direction or a "take" prompt). For "take", the plan is replaced with take action under the assumption that this is the last action to be performed. Otherwise, the plan is filled with actions that align with the direction prompt.

Then, the next action is chosen as described under on_silence.

- on_reference: The follower updates its internal target descriptor (color, shape, position) based on the new reference. Given this updated descriptor, the follower identifies candidate coordinates in the symbolic representation of the current field of view, for example, coordinates that are blue given a reference "Take the blue piece". If such a coordinate is identified and the follower has not already approached it, then the shortest path to that candidate is established as a new plan. Otherwise, if the descriptor only contains a position, then a direction towards that position is approached. In the case where the follower is already in that position, a randomly chosen piece in the field of view is approached. When none of this matches, then the current plan proceeds as described under on_silence.

Now, the autonomy defines which procedures the follower undertakes, when intermediate feedback *is missing* (the guide stays silent). The **cautious** follower is performing solely the previously defined procedures: when the plan is exhausted, then it waits until a new directive or reference is given. If this follower is over an assumed target piece, then it waits until the "take" directive is given by the guide. In contrast, the **eager** follower aims to actually take an assumed target piece when approaching it in the current field of view. Furthermore, the eager follower autonomously looks for target candidates at each step (as described in the on_reference procedure) and potentially revises the plan (also when the guide stays silent).

5 Learning Communication Policies for Different Follower Behaviors

Mnih et al. (2015) showed that vision-driven reinforcement learning policies can achieve human-level performance in pixel-based environments like Atari games. Similarly, the guide as an agent in our environment has the challenging task to learn:

(a) when to produce an utterance (or stay silent),
(b) what to produce (confirm, decline, direct, refer), and
(c) how to produce it (which directive or preference order)

based solely on visual observation of the board state and the follower actions.

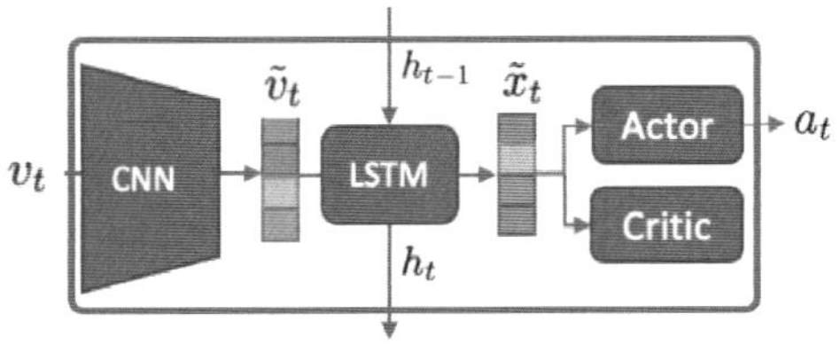

Figure 3: The guide's recurrent vision network.

5.1 The Guide

The observation $o_t = (v_t)$ with $v_t \in \mathbb{R}^{21\times21\times5}$ is encoded into a 128-dimensional feature vector $\tilde{v}_t \in \mathbb{R}$ using a 4-layer convolutional neural network similar to that by Chevalier-Boisvert et al. (2019). Then, the feature vector $\tilde{v}_t$ is fed through an LSTM (Hochreiter and Schmidhuber, 1997) which functions as a memory mechanism (updating a state vector h_t that is passed forward in time). Given the resulting memory-conditioned visual feature vector $\tilde{x}_t$, we learn a parameterized actor-critic-based policy $\pi(\tilde{x}_t; \theta) \sim a_t$ where the actor predicts a distribution over the action space (intents) and the critic estimates the value of the current state (Figure 3). For the recurrent policy, we use the implementation of *StableBaselines3-Contrib* v1.8.0 (Raffin et al., 2021), which performs back-propagation through time until the first step in an episode.

5.2 Experiment Setup

We employ the Proximal Policy Optimization (PPO) algorithm (Schulman et al., 2017) for policy learning in our sparse reward environment that respects an assumed accumulated effort over actions. Then we evaluate to which extent the resulting policies (the guides) are adapted towards the follower behaviors in such ways that align with expectations based on the follower's dimensions of confidence and autonomy. Thus, for the experiments, we initiate different **cautious** and **eager** follower's with increasing confidence discount factors so that $\phi \in [0.75, 0.85, 0.90, 0.95, 0.97, 0.99]$.

We use *StableBaselines3* v1.8.0 (Raffin et al., 2021) to learn for each of these follower behaviors a separate guide. We train each guide with 4 parallel running environments (batch size) and 1 million time steps in total. This means that each board in the training split is seen at least 13 times. Every 100k steps during training, we evaluate the pairings against the validation set. We keep for each pairing the guides that achieve the highest mean episode reward based on these validation runs. We conduct the experiments with three different seeds.

5.3 Results and Discussion

Overall Results. The overall results in Table 2 show that learned policies are communicative strategies that can successfully guide the follower (towards the target piece) in most of the cases (on average in 92% of the test episodes). This indicates that the guide learned the goal of the game and hereby almost reaches the best episode length (on average only 1.93 steps longer than the shortest path). The overall average effort (9.72) covers only about 71.5% of the average episode length (13.58) which means that the policies altogether produce an utterance in about 2 out of 3 steps.

Has the guide learned to stay silent? Indeed, Figure 4 shows that the policies converge to a mode where the `silence` intent is chosen in at least 23% of the steps: The guides are in general able to learn to say nothing. The most chosen intent is `reference` which is reasonable because it provides crucial information (the target piece description) and triggers an update of the follower's plan.

What preference orders are chosen for the `reference` production? The `reference` intents define the order in which properties are compared between the target piece and its distractors. This means, for example, that a CSP reference is likely to mention the target piece's color because the color attribute is first compared to distinguish the target from its distractors (and it is very likely that at least one distractor gets excluded because otherwise, all pieces would share the same color). Thus, it is reasonable that there are communicative strategies learned that choose CSP in the majority of cases as shown in Figure 5. This means that the guide produces a reference that likely includes the shape and the color of the target piece. These properties are indeed useful for the follower to identify and approach the target in its field of view. On the other hand, preference orders that test positions first (PCS and PSC) are also chosen rather often. These strategies lead the follower to the target piece without having it necessarily already in the field of view.

The effects of the follower's autonomy mode. We experimented with two levels of autonomy of the follower. The results in Table 2 show that the policies that learn from interactions with the **eager** follower require on average 2.00 points less effort than the **cautious** one. This is reasonable as the eager follower is autonomously updating the plan and looking for target candidates at each step. Along

Metrics:	mR ↑	mSR ↑	mEPL ↓	mEff. ↓
— Cautious —				
100% Silent	0.00	0.00	30.00	0.00
100% Ref.	-1.04	0.00	30.00	34.8
PPO-Guide	1.55	0.94	13.97	10.72
ϕ=75	1.52	0.93	15.02	11.07
ϕ=85	1.47	**0.96**	14.13	14.63
ϕ=90	1.59	0.95	13.87	10.33
ϕ=95	1.57	0.94	13.67	10.49
ϕ=97	1.57	0.93	13.27	10.00
ϕ=99	1.57	0.90	13.88	7.78
— Eager —				
100% Silent	0.45	0.23	16.78	0.00
100% Ref.	0.86	0.75	18.57	21.09
PPO-Guide	1.57	0.91	13.19	8.72
ϕ=75	1.54	0.92	13.54	10.04
ϕ=85	**1.60**	0.89	14.28	**6.15**
ϕ=90	1.49	0.92	13.24	11.67
ϕ=95	1.59	0.92	12.86	8.39
ϕ=97	1.58	0.90	12.64	7.28
ϕ=99	1.59	0.93	**12.58**	8.76
— Overall —				
100% Silent	0.23	0.11	23.39	0.00
100% Ref.	-0.09	0.37	24.29	27.94
PPO-Guide	1.56	0.92	13.58	9.72

Table 2: The mean rewards (mR), success rates (mSR in %), episodes lengths (mEPL) and efforts of the agents on the test tasks for the chosen autonomy and confidence combinations of the follower (averaged over all seeds). A shortest path solver reaches 11.65 mEPL (3.13 std). Given this, the upper bound for the mean reward is 1.83. Best values in bold.

Chosen Intent:	S	C	D	O	R
— Cautious —					
PPO-Guide	0.27	0.04	/	0.09	0.60
ϕ=75	0.27	0.08	/	0.08	0.56
ϕ=85	0.06	0.08	/	0.09	0.78
ϕ=90	0.29	0.09	/	0.08	0.53
ϕ=95	0.28	/	/	0.09	0.63
ϕ=97	0.30	/	/	0.09	0.61
ϕ=99	0.43	/	/	0.09	0.48
— Eager —					
PPO-Guide	0.34	0.06	0.06	0.09	0.46
ϕ=75	0.25	0.26	0.03	0.08	0.38
ϕ=85	0.53	0.01	0.09	0.08	0.29
ϕ=90	0.16	0.05	0.11	0.08	0.59
ϕ=95	0.34	/	0.13	0.09	0.45
ϕ=97	0.42	/	/	0.11	0.47
ϕ=99	0.33	0.02	/	0.08	0.57
— Overall —					
PPO-Guide	0.31	0.05	0.03	0.09	0.53

Table 3: The intent's mean chance of being chosen at a step (for each policy evaluated on the test split) broken down by a follower's confidence and autonomy. The intents are abbreviated as follows: `silence` (S), `confirm` (C), `decline` (D), `directive` (O) and `reference` (R). It appears reasonable that the cautious follower's actions are never declined because the behavior is to always wait for the guide's instructions (in contrast to the eager ones that explore occasionally on their own). Similarly, the higher confidence follower's require less re-assurance (confirms) of their actions.

these lines, it is also reasonable that the decline intent is never selected for the cautious follower (see Table 3) because it never tried to approach a target piece without the guide referencing it.

The effects of the follower's confidence. The differences in the intent selection strategy of the learned policies (guides) shown in Table 3 indicate that guides learned from interaction with more confident follower's ($\phi > 0.9$) produce less or no `confirm` actions. This seems reasonable as the decrease in the execution probability of these followers is less steep and a `reference` action has a similar effect. Furthermore, we see a slight tendency of guides to stay quieter (on average) when trained with more confident followers as shown in Figure 6. However we cannot see such a tendency for guides trained with less confident followers.

6 Conclusions

In this work, we examined an interesting intersection between psycho-linguistic studies and deep learning with reinforcement learning. We considered neural agents as possible interaction partners (for humans) in a challenging reference game where a guide has to learn when, what, and how information (actionable intents) is to be provided to a follower. As a proxy for different follower behaviors, we implemented a hand-crafted policy that is controllable along two dimensions: autonomy in exploration and confidence in executing an action. We experimented with a learning signal that in addition to the goal condition also respects an assumed communicative effort. Our results indicate that this formulation of the learning signal leads to communicative strategies that are less verbose (stay silent more often) and that the resulting guide behaviors are adapted (in terms of intent selection distributions) to the follower's autonomy and confidence levels. We think this work presents a useful case study of neural agents that have to learn adapted communication strategies in an interactive setting (possibly with humans). In future work, we want to investigate other reward formulations for the reference game and evaluate the learning of communication policies where the utterance production process spans multiple time steps (one word at a time) and the production must be possibly interrupted and revised during the interaction.

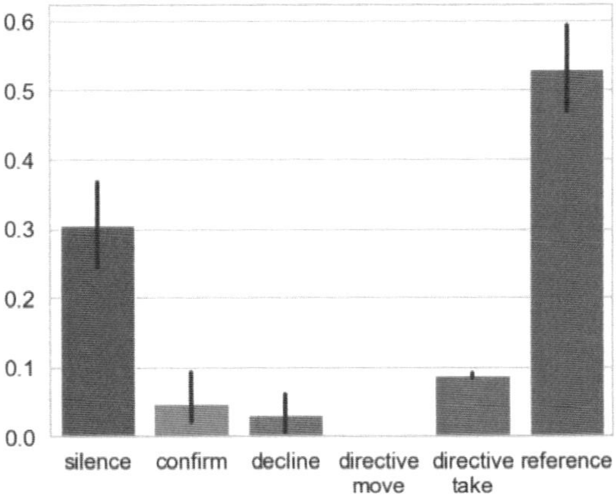

Figure 4: An intent's mean chance of being chosen at a step (for all learnt policies evaluated on the test split).

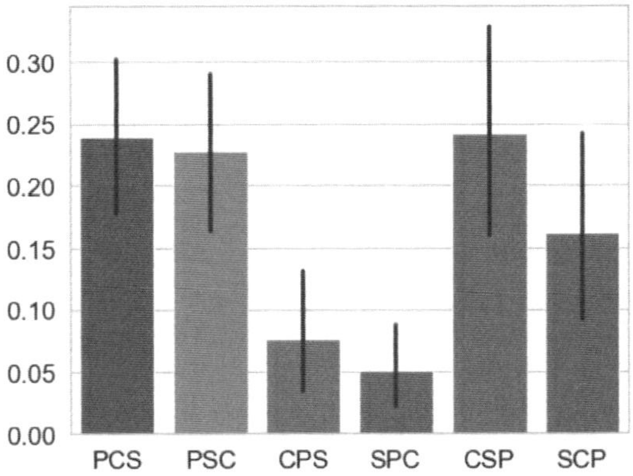

Figure 5: The distribution of the preference order choices for the `reference` action (from Figure 4). The preferences over position (P), shape (S) and color (C) are given to the IA for reference production.

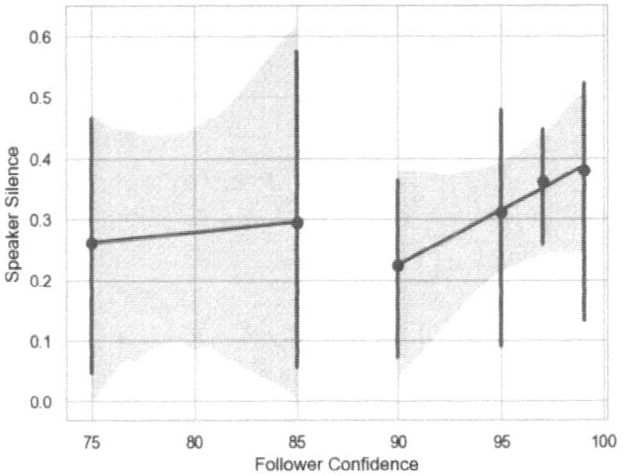

Figure 6: The mean number of silent turns performed by the learnt policies (incl. all seeds) during the test episodes. We fitted a linear regression with a confidence interval of 99% through the data points separately for the followers with $\phi = \{75, 85\}$ and $\phi = \{90, 95, 97, 99\}$. The latter shows a trend towards more silence turns when the guide is paired with more confident followers.

Acknowledgements

We want to thank the anonymous reviewers for their comments. This work was funded by the *Deutsche Forschungsgemeinschaft* (DFG, German Research Foundation) – 423217434 ("RECOLAGE") grant.

References

Stefano V. Albrecht and Peter Stone. 2018. Autonomous agents modelling other agents: A comprehensive survey and open problems. *Artif. Intell.*, 258:66–95.

Elliot Chane-Sane, Cordelia Schmid, and Ivan Laptev. 2021. Goal-conditioned reinforcement learning with imagined subgoals. In *Proceedings of the 38th International Conference on Machine Learning, ICML 2021, 18-24 July 2021, Virtual Event*, volume 139 of *Proceedings of Machine Learning Research*, pages 1430–1440. PMLR.

Can Chang, Ni Mu, Jiajun Wu, Ling Pan, and Huazhe Xu. 2022. E-MAPP: efficient multi-agent reinforcement learning with parallel program guidance. In *Advances in Neural Information Processing Systems 35: Annual Conference on Neural Information Processing Systems 2022, NeurIPS 2022, New Orleans, LA, USA, November 28 - December 9, 2022*.

Maxime Chevalier-Boisvert, Dzmitry Bahdanau, Salem Lahlou, Lucas Willems, Chitwan Saharia, Thien Huu Nguyen, and Yoshua Bengio. 2019. Babyai: A platform to study the sample efficiency of grounded language learning. In *7th International Conference on Learning Representations, ICLR 2019, New Orleans, LA, USA, May 6-9, 2019*. OpenReview.net.

Herbert H. Clark and Deanna Wilkes-Gibbs. 1986. Referring as a collaborative process. *Cognition*, 22(1):1–39. Place: Netherlands Publisher: Elsevier Science.

John D. Co-Reyes, Abhishek Gupta, Suvansh Sanjeev, Nick Altieri, Jacob Andreas, John DeNero, Pieter Abbeel, and Sergey Levine. 2019. Guiding policies with language via meta-learning. In *7th International Conference on Learning Representations, ICLR 2019, New Orleans, LA, USA, May 6-9, 2019*. OpenReview.net.

Robert Dale and Ehud Reiter. 1995. Computational interpretations of the gricean maxims in the generation of referring expressions. *Cogn. Sci.*, 19(2):233–263.

Ido Davidesco, Emma Laurent, Henry Valk, Tessa West, Catherine Milne, David Poeppel, and Suzanne Dikker. 2023. The Temporal Dynamics of Brain-to-Brain Synchrony Between Students and Teachers Predict Learning Outcomes. *Psychological Science*, 34(5):633–643.

Haihong E, Peiqing Niu, Zhongfu Chen, and Meina Song. 2019. A novel bi-directional interrelated model for joint intent detection and slot filling. In *Proceedings of the 57th Conference of the Association for Computational Linguistics, ACL 2019, Florence, Italy, July 28- August 2, 2019, Volume 1: Long Papers*, pages 5467–5471. Association for Computational Linguistics.

Daniel Fried, Ronghang Hu, Volkan Cirik, Anna Rohrbach, Jacob Andreas, Louis-Philippe Morency, Taylor Berg-Kirkpatrick, Kate Saenko, Dan Klein, and Trevor Darrell. 2018. Speaker-follower models for vision-and-language navigation. In *Advances in Neural Information Processing Systems 31: Annual Conference on Neural Information Processing Systems 2018, NeurIPS 2018, December 3-8, 2018, Montréal, Canada*, pages 3318–3329.

Xiaofeng Gao, Qiaozi Gao, Ran Gong, Kaixiang Lin, Govind Thattai, and Gaurav S. Sukhatme. 2022. Dialfred: Dialogue-enabled agents for embodied instruction following. *IEEE Robotics Autom. Lett.*, 7(4):10049–10056.

Ahana Ghosh, Sebastian Tschiatschek, Hamed Mahdavi, and Adish Singla. 2020. Towards deployment of robust cooperative AI agents: An algorithmic framework for learning adaptive policies. In *Proceedings of the 19th International Conference on Autonomous Agents and Multiagent Systems, AAMAS '20, Auckland, New Zealand, May 9-13, 2020*, pages 447–455. International Foundation for Autonomous Agents and Multiagent Systems.

Solomon W. Golomb. 1996. *Polyominoes: Puzzles, Patterns, Problems, and Packings*. Princeton University Press.

Jing Gu, Eliana Stefani, Qi Wu, Jesse Thomason, and Xin Wang. 2022. Vision-and-language navigation: A survey of tasks, methods, and future directions. In *Proceedings of the 60th Annual Meeting of the Association for Computational Linguistics (Volume 1: Long Papers), ACL 2022, Dublin, Ireland, May 22-27, 2022*, pages 7606–7623. Association for Computational Linguistics.

Nico Gürtler, Dieter Büchler, and Georg Martius. 2021. Hierarchical reinforcement learning with timed subgoals. In *Advances in Neural Information Processing Systems 34: Annual Conference on Neural Information Processing Systems 2021, NeurIPS 2021, December 6-14, 2021, virtual*, pages 21732–21743.

Sepp Hochreiter and Jürgen Schmidhuber. 1997. Long Short-Term Memory. *Neural Computation*, 9(8):1735–1780.

Bin Hu, Chenyang Zhao, Pu Zhang, Zihao Zhou, Yuanhang Yang, Zenglin Xu, and Bin Liu. 2023. Enabling intelligent interactions between an agent and an LLM: A reinforcement learning approach. *CoRR*, abs/2306.03604.

Sukai Huang, Nir Lipovetzky, and Trevor Cohn. 2023. A reminder of its brittleness: Language reward shaping may hinder learning for instruction following agents. *CoRR*, abs/2305.16621.

Jeewon Jeon, Woojun Kim, Whiyoung Jung, and Youngchul Sung. 2022. MASER: multi-agent reinforcement learning with subgoals generated from experience replay buffer. In *International Conference on Machine Learning, ICML 2022, 17-23 July 2022, Baltimore, Maryland, USA*, volume 162 of *Proceedings of Machine Learning Research*, pages 10041–10052. PMLR.

Tom Jurgenson and Aviv Tamar. 2023. Goal-conditioned supervised learning with sub-goal prediction. *CoRR*, abs/2305.10171.

Gyeong Taek Lee and Kang Jin Kim. 2023. A controllable agent by subgoals in path planning using goal-conditioned reinforcement learning. *IEEE Access*, 11:33812–33825.

Yuntao Liu, Yuan Li, Xinhai Xu, Yong Dou, and Donghong Liu. 2022. Heterogeneous skill learning for multi-agent tasks. In *Advances in Neural Information Processing Systems 35: Annual Conference on Neural Information Processing Systems 2022, NeurIPS 2022, New Orleans, LA, USA, November 28 - December 9, 2022*.

Ryan Lowe, Jakob N. Foerster, Y-Lan Boureau, Joelle Pineau, and Yann N. Dauphin. 2019. On the pitfalls of measuring emergent communication. In *Proceedings of the 18th International Conference on Autonomous Agents and MultiAgent Systems, AAMAS '19, Montreal, QC, Canada, May 13-17, 2019*, pages 693–701. International Foundation for Autonomous Agents and Multiagent Systems.

Volodymyr Mnih, Koray Kavukcuoglu, David Silver, Andrei A. Rusu, Joel Veness, Marc G. Bellemare, Alex Graves, Martin A. Riedmiller, Andreas Fidjeland, Georg Ostrovski, Stig Petersen, Charles Beattie, Amir Sadik, Ioannis Antonoglou, Helen King, Dharshan Kumaran, Daan Wierstra, Shane Legg, and Demis Hassabis. 2015. Human-level control through deep reinforcement learning. *Nat.*, 518(7540):529–533.

Seungwhan Moon, Satwik Kottur, Paul A. Crook, Ankita De, Shivani Poddar, Theodore Levin, David Whitney, Daniel Difranco, Ahmad Beirami, Eunjoon Cho, Rajen Subba, and Alborz Geramifard. 2020. Situated and interactive multimodal conversations. In *Proceedings of the 28th International Conference on Computational Linguistics, COLING 2020, Barcelona, Spain (Online), December 8-13, 2020*, pages 1103–1121. International Committee on Computational Linguistics.

Jesse Mu, Victor Zhong, Roberta Raileanu, Minqi Jiang, Noah D. Goodman, Tim Rocktäschel, and Edward Grefenstette. 2022. Improving intrinsic exploration with language abstractions. In *Advances in Neural Information Processing Systems 35: Annual Conference on Neural Information Processing Systems 2022, NeurIPS 2022, New Orleans, LA, USA, November 28 - December 9, 2022*.

Khanh Nguyen and Hal Daumé III. 2019. Help, anna! visual navigation with natural multimodal assistance via retrospective curiosity-encouraging imitation learning. In *Proceedings of the 2019 Conference on Empirical Methods in Natural Language Processing and the 9th International Joint Conference on Natural Language Processing, EMNLP-IJCNLP 2019, Hong Kong, China, November 3-7, 2019*, pages 684–695. Association for Computational Linguistics.

Khanh Nguyen, Debadeepta Dey, Chris Brockett, and Bill Dolan. 2019. Vision-based navigation with language-based assistance via imitation learning with indirect intervention. In *IEEE Conference on Computer Vision and Pattern Recognition, CVPR 2019, Long Beach, CA, USA, June 16-20, 2019*, pages 12527–12537. Computer Vision Foundation / IEEE.

Aishwarya Padmakumar, Jesse Thomason, Ayush Shrivastava, Patrick Lange, Anjali Narayan-Chen, Spandana Gella, Robinson Piramuthu, Gökhan Tür, and Dilek Hakkani-Tür. 2022. Teach: Task-driven embodied agents that chat. In *Thirty-Sixth AAAI Conference on Artificial Intelligence, AAAI 2022, Thirty-Fourth Conference on Innovative Applications of Artificial Intelligence, IAAI 2022, The Twelveth Symposium on Educational Advances in Artificial Intelligence, EAAI 2022 Virtual Event, February 22 - March 1, 2022*, pages 2017–2025. AAAI Press.

Alexander Pashevich, Cordelia Schmid, and Chen Sun. 2021. Episodic transformer for vision-and-language navigation. In *2021 IEEE/CVF International Conference on Computer Vision, ICCV 2021, Montreal, QC, Canada, October 10-17, 2021*, pages 15922–15932. IEEE.

Adam Paszke, Sam Gross, Francisco Massa, Adam Lerer, James Bradbury, Gregory Chanan, Trevor Killeen, Zeming Lin, Natalia Gimelshein, Luca Antiga, Alban Desmaison, Andreas Köpf, Edward Z. Yang, Zachary DeVito, Martin Raison, Alykhan Tejani, Sasank Chilamkurthy, Benoit Steiner, Lu Fang, Junjie Bai, and Soumith Chintala. 2019. Pytorch: An imperative style, high-performance deep learning library. In *Advances in Neural Information Processing Systems 32: Annual Conference on Neural Information Processing Systems 2019, NeurIPS 2019, December 8-14, 2019, Vancouver, BC, Canada*, pages 8024–8035.

Karl Pertsch, Oleh Rybkin, Frederik Ebert, Shenghao Zhou, Dinesh Jayaraman, Chelsea Finn, and Sergey Levine. 2020. Long-horizon visual planning with goal-conditioned hierarchical predictors. In *Advances in Neural Information Processing Systems 33: Annual Conference on Neural Information Processing Systems 2020, NeurIPS 2020, December 6-12, 2020, virtual*.

Antonin Raffin, Ashley Hill, Adam Gleave, Anssi Kanervisto, Maximilian Ernestus, and Noah Dormann. 2021. Stable-baselines3: Reliable reinforcement learning implementations. *Journal of Machine Learning Research*, 22(268):1–8.

Philipp Sadler, Sherzod Hakimov, and David Schlangen. 2023. Yes, this way! learning to ground referring expressions into actions with intra-episodic feedback from supportive teachers. In *Findings of the Association for Computational Linguistics: ACL 2023, Toronto, Canada, July 9-14, 2023*, pages 9228–9239. Association for Computational Linguistics.

John Schulman, Filip Wolski, Prafulla Dhariwal, Alec Radford, and Oleg Klimov. 2017. Proximal policy optimization algorithms. *CoRR*, abs/1707.06347.

Alane Suhr and Yoav Artzi. 2023. Continual learning for instruction following from realtime feedback. In *Advances in Neural Information Processing Systems 36: Annual Conference on Neural Information Processing Systems 2023, NeurIPS 2023, New Orleans, LA, USA, December 10 - 16, 2023*.

Haotian Sun, Yuchen Zhuang, Lingkai Kong, Bo Dai, and Chao Zhang. 2023. Adaplanner: Adaptive planning from feedback with language models. In *Advances in Neural Information Processing Systems 36: Annual Conference on Neural Information Processing Systems 2023, NeurIPS 2023, New Orleans, LA, USA, December 10 - 16, 2023*.

Richard S. Sutton and Andrew G. Barto. 2018. *Reinforcement Learning: An Introduction*, second edition. The MIT Press.

Jesse Thomason, Michael Murray, Maya Cakmak, and Luke Zettlemoyer. 2019. Vision-and-dialog navigation. In *3rd Annual Conference on Robot Learning, CoRL 2019, Osaka, Japan, October 30 - November 1, 2019, Proceedings*, volume 100 of *Proceedings of Machine Learning Research*, pages 394–406. PMLR.

Kees van Deemter. 2016. *Computational Models of Referring*, chapter 4.6. The MIT Press.

Woodrow Zhouyuan Wang, Andy Shih, Annie Xie, and Dorsa Sadigh. 2021. Influencing towards stable multi-agent interactions. In *Conference on Robot Learning, 8-11 November 2021, London, UK*, volume 164 of *Proceedings of Machine Learning Research*, pages 1132–1143. PMLR.

Annie Xie, Dylan P. Losey, Ryan Tolsma, Chelsea Finn, and Dorsa Sadigh. 2020. Learning latent representations to influence multi-agent interaction. In *4th Conference on Robot Learning, CoRL 2020, 16-18 November 2020, Virtual Event / Cambridge, MA, USA*, volume 155 of *Proceedings of Machine Learning Research*, pages 575–588. PMLR.

Mingyu Yang, Jian Zhao, Xunhan Hu, Wengang Zhou, Jiangcheng Zhu, and Houqiang Li. 2022. LDSA: learning dynamic subtask assignment in cooperative multi-agent reinforcement learning. In *Advances in Neural Information Processing Systems 35: Annual Conference on Neural Information Processing Systems 2022, NeurIPS 2022, New Orleans, LA, USA, November 28 - December 9, 2022*.

Sina Zarrieß, Julian Hough, Casey Kennington, Ramesh Manuvinakurike, David DeVault, Raquel Fernández, and David Schlangen. 2016. PentoRef: A Corpus of Spoken References in Task-oriented Dialogues. In *Proceedings of the Tenth International Conference on Language Resources and Evaluation (LREC'16)*, pages 125–131, Portorož, Slovenia. European Language Resources Association (ELRA).

A Appendix

Robot image in Figure 1 adjusted from `https://commons.wikimedia.org/wiki/File:Cartoon_Robot.svg`. That file was made available under the Creative Commons CC0 1.0 Universal Public Domain Dedication.

A.1 Environment Details

Board The internal representation of the visual state is a 2-dimensional grid that spans $W \times H$ tiles where W and H are defined by the map size. A tile is either empty or holds an identifier for a piece (the tile is then occupied). The pieces are defined by their colour, shape and coordinates and occupy five adjacent tiles (within a virtual box of 5×5 tiles). The pieces are not allowed to overlap with another piece's tiles. For a higher visual variation, we also apply rotations to pieces, but we ignore the rotation for expression generation, though this could be an extension of the task. The colors are described in Table 4.

Name	HEX	RGB
red	#ff0000	(255, 0, 0)
green	#008000	(0, 128, 0)
blue	#0000ff	(0, 0, 255)
yellow	#ffff00	(255, 255, 0)
brown	#8b4513	(139, 69, 19)
purple	#800080	(128, 0, 128)

Table 4: The colors for the Pentomino pieces.

Symbols The symbolic repesentations for the shapes are: P (2), X (3), T (4), Z (5), W (6), U (7), N (8), F (9), Y (10). The colors are encoded as: red (2), green (3), blue (4), yellow (5), brown (6), purple (7). The 0-symbol is reserved for out-of-world tiles (which can occur in the partial view). The 1-symbol is reserved for an empty tile.

Gripper The gripper can only move one position at a step and can move over pieces, but is not allowed to leave the boundaries of the board. The gripper coordinates are defined as $\{(x, y) : x \in [0, W], y \in [0, H]\}$.

The IA on symbolic properties as based on the formulation by van Deemter (2016)

Require: A set of distractors M, a set of property values $\mathcal{P}$ of a referent r and a linear preference order $\mathcal{O}$ over the property values $\mathcal{P}$

```
D ← ∅
for P in O(P) do
    E ← {m ∈ M : ¬P(m)}
    if E ≠ ∅ then
        Add P to D
        Remove E from M
    return D
```

References The Incremental Algorithm (Algorithm 1), in the formulation of (Dale and Reiter, 1995), is supposed to find the properties that uniquely identify an object among others given a preference over properties. To accomplish this the algorithm is given the property values $\mathcal{P}$ of distractors in M and of a referent r. Then the algorithm excludes distractors in several iterations until either M is empty or every property of r has been tested. During the exclusion process the algorithm computes the set of distractors that do *not* share a given property with the referent and stores the property in $\mathcal{D}$. These properties in $\mathcal{D}$ are the ones that distinguish the referent from the others and thus will be returned.

The algorithm has a meta-parameter $\mathcal{O}$, indicating the *preference order*, which determines the order in which the properties of the referent are tested against the distractors. In our domain, for example, when *color* is the most preferred property, the algorithm might return BLUE, if this property already excludes all distractors. When *shape* is the preferred property and all distractors do *not* share the shape T with the referent, T would be returned. Hence even when the referent and distractor pieces are the same, different preference orders might lead to different expressions.

There are 3 expression templates that are used when only a single property value of the target piece is returned by the Incremental Algorithm (IA):

- *Take the [color] piece*
- *Take the [shape]*
- *Take the piece at [position]*

Then there are 3 expression templates that are selected when two properties are returned:

- *Take the [color] [shape]*
- *Take the [color] piece at [position]*
- *Take the [shape] at [position]*

And finally there is one expression templates that lists all property values to identify a target piece:

- *Take the [color] [shape] at [position]*

Vocabulary Overall, the property values and sentence templates lead to a small vocabulary of 37 words:

- 9 shapes: P, X, T, Z, W, U, N, F, Y
- 6 colors: red, green, blue, yellow, brown, purple
- 6 position words: left, right, top, bottom, center (which are combined to e.g., right center or top left)
- 12 template words: take, the, piece, at, yes, no, this, way, go, a, bit, more
- 4 special words: <s>, <e>, <pad>, <unk>

The maximal sentence length is 12.

A.2 Task Details

To create a task, we first place the target piece on a board. Then, we sample uniformly random from all possible pieces and place them until the wanted number of pieces is reached (we experiment with 2 to 8 pieces on a board). If a piece cannot be placed after a certain amount of tries, then we resample a piece and try again. The coordinates are chosen at random uniform from the coordinates that fall into an area of the symbolic description. We never set a piece into the center, because that is the location where the gripper is initially located. In this way, we construct 100 training boards (or 1 evaluation board respectively) for each number of pieces (2-8). To ensure that a board scene in the training split cannot be aligned with a target piece, we create 3 extra tasks for a single board by choosing extra targets (when fewer than 4 pieces are on a board, then we create a task for each piece). For evaluation, we only create a single task for each target piece symbol.

A.3 Guide Details

Agent Parameters: $602,447$

feature_dims	128
normalize_images	True
shared_lstm	True
enable_critic_lstm	False
n_lstm_layers	1
lstm_hidden_size	128

Table 5: Policy arguments for the the RecurrentPPO agent

Policy Architecture We instantiate the actor-critic PPO agent with an architecture defined by `pi=[64, 64], vf=[64, 64]` meaning that the actor is a 2-layer feedforward network with 64 parameters per layer. The critic has the same architecture, but does not share the weights with the actor.

Vision Encoder The visual encoder is a convolutional neural network (CNN) with 4 layers that maps the visual observations $v_t \in \mathbb{R}^{21\times21\times5}$ into a 128-dimensional features vector $\tilde{v} \in \mathbb{R}$. We consecutively apply four blocks of `(nn.Conv2d(),nn.BatchNorm2d(),nn.ReLU())` with same padding where the kernel size is 3×3, except for the first blocke where we set the kernel size to 1×1. After the fourth block we apply a `nn.AdaptiveMaxPool2d((1, 1))` layer from PyTorch v1.13.0 (Paszke et al., 2019) to collapse the spatial dimensions of the feature maps.

Learning Algorithm We use the RecurrentPPO implementation from StableBaselines-Contrib v1.8.0 (Raffin et al., 2021) with the hyper-parameters in Table 6 (and the defaults otherwise).

learning_rate	3e-4
clip_range	0.2
gamma	0.99
gae_lambda	0.95
ent_coef	0.0
vf_coef	0.5
max_grad_norm	0.5
lr_init	3e-4
n_steps	128
batch_size	128
num_epochs	10

Table 6: RecurrentPPO hyperparameters

A.4 Experiment Details

We trained the agents simultaneously on 8 GeForce GTX 1080 Ti (11GB) where each of them consumed about 4GB of GPU memory. The training for the 36 configurations took around 144 hours in total (about $4h$ for the 1 million steps each). The random seeds were set to 49184, 98506 or 92999 respectively. As the evaluation criteria on the testings tasks we chose success rate which indicates the relative number of episodes (in a rollout or in a test split) where the agent selected the correct piece:

$$\text{mSR} = \frac{\sum^N s_i}{N} \text{ where } s_i = \begin{cases} 1, & \text{for correct piece} \\ 0, & \text{otherwise} \end{cases}$$

Efforts. We choose $E_{\text{Guide}} := \{0, 1.0, 1.1, 1.2\}$ for the efforts of the categories of actionable intent in such a way that the silence action is the one with the least effort. The silence action simply results into the metabolic costs necessary to perform the task over multiple time steps (the game reward). The other language actions introduce an additional effort. These actions should differ on the magnitude in such a way that they can be ordered based on the effort where the reference production is presumably taking the most effort (1.2) and a confirmation ("Yes") or rejection signal ("No") is taking less effort (1.0). We assumed that directive are on the middle ground (1.1) and that they should appear more often, when used. This basically means for around every 10th action an additional non-silence action can be taken, when choosing to use directives over references. Moreover, when using the maximal number of 30 steps and only taking the respective actions, this results into an effort reward of $1 - (0.9 \cdot 1.2) = -0.08$ (slightly negative) for the references and $1 - (0.9 \cdot 1.1) = 0.01$ (slightly positive) for the directives and $1 - (0.9 \cdot 1.0) = 0.1$ (still positive) for the confirmations or corrections. These magnitudes are supposed to be close to the initial formulation for the game reward and thus around -2 and $+2$ (incl. the outcome) to keep the learning of the value function more stable. We note that the signal for the ordering of the actionable intents is very small, but it should make an effect.

Collection of Japanese Route Information Reference Expressions Using Maps as Stimuli

Yoshiko Kawabata[1], Mai Omura[1], Hikari Konishi[2], Masayuki Asahara[1], Johane Takeuchi[3]
[1]National Institute for Japanese Language and Linguistics, Japan,
[2]Tecca LLC,
[3]Honda Research Institute
Correspondence: masayu-a@ninjal.ac.jp

Abstract

We constructed a database of Japanese expressions based on route information for language-based direction instructions to autonomous driving systems. Using 20 maps as stimuli, we requested descriptions of routes between two points on each map from 40 individuals per route, collecting 1600 route information reference expressions. We determined whether the expressions were based solely on relative reference expressions by using landmarks on the maps. In cases in which only relative reference expressions were used, we labeled the presence or absence of information regarding the starting point, waypoints, and destination. Additionally, we collected clarity ratings for each expression using a survey.

1 Introduction

Accurately conveying route information in a language is challenging because it comprises details regarding the starting point, waypoints, and destination. Utilizing surrounding landmarks is crucial for effectively conveying positional information, movement direction, and distance. In languages where case elements such as subjects tend to be omitted (e.g., Japanese), it is difficult to generate clear expressions of route information.

By collecting Japanese route information reference expressions using maps as stimuli, this study aims to shed light on important perspectives for conveying route information for language-based direction instructions to autonomous driving systems. For each of the 20 maps, two starting and ending point patterns were established, resulting in 40 stimuli each. Through crowdsourcing, we sought to gather 40 expressions of route information references per map, articulated solely using specific or relative location information. Following the determination of whether the collected language expressions comprised only specific or relative location information, we annotated the inclusion of starting point, waypoint, or destination information. Furthermore, the clarity ratings for each expression were obtained through surveys.

2 Related Research

Early work on the analysis of direction-giving conversations was conducted by Psathas and Kozloff (1976), who identified three stages: situating, specifying information and directions, and concluding. They pointed out that during the initial directions, a) starting point, b) destination, c) mode of travel, d) time of travel, and e) membership categorization by the parties of each other are important. In a related study, Clark and Wilkes-Gibbs (1986) introduced a collaborative model to make definite references in conversations. In their model, speakers initiate the process by presenting a noun phrase, which is iteratively refined by participants until a mutually accepted version is reached, thus minimizing joint effort. In such dialogues, there is often a significant amount of coreference information regarding locations. Additionally, Levinson (2004) focused on the relationship between language and cognition and explored the role of space in cognitive diversity. Moreover, Lakoff (1987) proposed a SOURCE-PATH-GOAL Schema, identifying four structural elements: a) SOURCE, b) DESTINATION, c) PATH, and d) DIRECTION. Barclay and Galton (2008) constructed 'scene corpus' in which spatial expressions were collected by showing virtual objects. Shelton and McNamara (2004) performed three experiments based on the fictional environments described by Taylor and Tversky (1992). They evaluated the mental costs associated with switching between route and survey perspective stimuli. Their experiments utilized eight patterns of maps rotated at 45° increments. In contrast, this study collects route expressions by showing a real-world map instead of focusing on positions.

Proceedings of the 4th Workshop on Spatial Language Understanding and Grounded Communication for Robotics (SpLU-RoboNLP 2024), pages 30–35
August 11-16, 2024

3 Data Collection Methodology

3.1 Collection of Route Information Reference Expressions

Route information reference expressions were collected using Yahoo! Crowd Sourcing. Participants were provided with maps, as shown in Figure 1, and were asked to describe any route information starting from ■ and ending at ★ (for screening) or ● (for main task).

During the task, the goal was to collect relative reference expressions. Participants were presented with a rotated version of the original map as stimuli and given the following instructions:

- Use "front, back, left, and right" as if the participants were initially located at the ■ mark on the map.
- Do not use "up, down, left, and right."
- Do not use "east, west, south, and north."

Figure 1: Example map used as stimuli (rotated 30°)

Expression collection was conducted in two stages: a screening survey and main survey. In the screening survey, maps rotated at angles of 30°, 120°, 210°, and 300° were used, as shown in Figure 1. Data were collected from 400 participants for each map for a total of 1600 responses. An example survey screen is shown in Appendix Figure 2. Participants were compensated with 10 yen equivalent PayPay points per response for the screening survey. The screening survey was conducted from 08:01 November 2, 2023, to 03:40 November 3, 2023. Among the participants, those who rated the clarity of expressions in the subsequent survey with a score of 3.0 or higher (206 individuals) were selected for the main survey. A clarity assessment of the screening survey results was conducted from 08:06 November 17, 2023, to 09:55 November 17, 2023.

In the main survey, 40 stimuli were used, each consisting of 20 types of maps with two starting and ending point patterns. A total of 1600 expressions were collected from 10 participants for each of the 160 variations of stimuli, which included four types of rotations for the 40 stimuli. Participants were compensated with 50 yen equivalent PayPal points per response for the main survey. The main survey was conducted from 14:02 November 17, 2023, to 23:55 November 19, 2023. Because we use a real-world map, there are multiple routes to the starting point and destination. In this context, we also attempt to collect expressions regarding which target points on the map are easiest to explain.

3.2 Classification of Collected Expressions

All 1600 expressions collected in the main survey exclusively consisted specific location information expressions and relative location information expressions, without the use of absolute location information such as east, west, south, and north.

However, five expressions were identified as inappropriate route information reference expressions because they mistakenly recognized incorrect marks on the map as the starting and ending points.

Additionally, 29 expressions were identified as inappropriate route information reference expressions because they recognized the starting and ending points in reverse (labeled as "W").

Subsequently, the following classifications were assigned (examples are based on Figure 1 as stimuli):

- X: Detailed description of the starting point
- Y: Description of points along the route
- Z: Detailed description of the destination

For the determination of the starting point (X), if there was a clear indication of the starting point such as "facing [location]," "standing in front of [location]," "between [location A] and [location B]," or "leaving from [location]," it was classified as starting point present (X). Additionally, if the starting point was explicitly stated as "from current location" or "from ■," it was also classified as starting point present (X).

For the determination of the route (Y), the presence of verbs indicating movement such as "turn left," "turn right," "go straight," "turn," or "go around" was used.

For the determination of the destination (Z), if explicit words indicating the destination such as "goal," "destination," or "arrive" were identified, it was classified as destination present (Z). Even if explicit words were not present , they were classified as destination present (Z) if there was a specific description of the destination. However, if there was no specificity regarding the destination, such as "after a while, on the left side," "beyond the plaza," or "passed by," it was classified as destination absent.

These classifications were set as multi-labels.

3.3 Clarity Rating of Expressions

The clarity of expressions was assessed through a survey using Yahoo! Crowd Sourcing for 1600 expressions collected in both the screening and main surveys. For the screening survey, 887 individuals who provided expressions consisting only of specific and relative location information were recruited, with 216 participants providing ratings for the screening survey data and 605 participants providing ratings for the main survey data.

A survey screen example is shown in Appendix Figure 3. Seven expressions were randomly presented for each map and ratings were made on a 6-point scale from 0 (difficult to understand) to 5 (easy to understand). Participants were compensated with 2 yen equivalent PayPal points per response. The clarity rating survey for the main survey collected ratings from 35 individuals per expression. The main survey was conducted from 17:01 December 14, 2023, to 13:10 December 16, 2023.

The expression and impression rating collection for this study was approved by the ethical review board of the National Institute for Japanese Language and Linguistics.

4 Data Statistics

Table 1 presents the individual aggregations based on the presence or absence of each aspect (TRUE for presence and FALSE for absence) along with the average clarity for each aspect.

In cases where the starting point and destination were mistaken (W), they tended to be less clear than the correct ones. A t-test was conducted, assuming unequal variances between the two samples. The mean clarity rating for W=FALSE was 2.79, whereas that for W=TRUE was 2.07. The difference between the means was significant ($t = 10.11$, $p < 0.001$, two-tailed).

Table 1: Clarity by Aspect (Individual)

Misidentification (W)	W=FALSE	W=TRUE
Count	1571	29
Clarity (Average)	2.79	2.07
Starting Point (X)	X=FALSE	X=TRUE
Count	599	1001
Clarity (Average)	2.72	2.81
Route (Y)	Y =FALSE	Y=TRUE
Count	24	1576
Clarity (Average)	2.00	2.79
Destination (Z)	Z=FALSE	Z=TRUE
Count	51	1549
Clarity (Average)	2.43	2.79

Explanations for the starting point (X) were frequently omitted, with 37.5% (599/1600) of the expressions lacking such an explanation. Those without a starting-point explanation exhibited slightly lower clarity than those with an explanation. A t-test assuming unequal variances between the two samples showed a significant difference ($t = -4.54$, $p < 0.001$, two-tailed), with a mean clarity rating of 2.72 for X=FALSE and 2.82 for X=TRUE.

The absence of route explanations (Y=FALSE) comprised 1.5% (24/1600) of the expressions, with an average clarity rating of 2.00, which was notably lower than the average clarity rating of 2.79 for expressions with route explanations (Y=TRUE). A t-test with unequal variances revealed a significant difference between the two groups ($t = -8.94$, $p < 0.001$, two-tailed).

Similarly, the absence of explanations for the destination (Z) was observed in 3.2% (51/1600) of the responses. These expressions had a lower clarity average of 2.43 compared to those with a destination explanation (clarity average of 2.79). A t-test assuming unequal variances between the two samples revealed a significant difference ($t = -5.78$, $p < 0.001$, two-tailed), with a mean clarity rating of 2.43 for Z=FALSE and 2.79 for Z=TRUE.

Hence, it is evident that all three aspects - starting point, route, and destination - play crucial roles in effectively conveying route information. Additionally, it i s noteworthy that explanations for the starting point (X) are often omitted, with 37.5% (599/1600) of the expressions lacking a starting point explanation. This observation underscores the challenge of referencing a starting point, which contributes to the complexity of comprehensively conveying route information.

Table 2: Clarity by Aspect (Combinations)

W	X	Y	Z	Clarity	Count
F	F	F	T	1.75	8
F	F	T	F	2.42	25
F	F	T	T	2.76	558
F	T	F	F	1.93	2
F	T	F	T	2.21	12
F	T	T	F	2.52	22
F	T	T	T	2.85	944
T	F	T	T	2.07	8
T	T	F	T	1.87	2
T	T	T	F	2.20	2
T	T	T	T	2.08	17
Total				2.78	1600

Table 2 presents the aggregation results based on the combination of aspects. From these combinations, it was observed that both route (Y) and destination (Z) explanations were predominant, accounting for 93.9% ((558+944)/1600). There was a difference in clarity depending on the presence of starting point (X) explanations; those with a starting point explanation (clarity average of 2.85) were clearer than those without (clarity average of 2.76).

Appendix Figure 4 shows one of clearest examples. This example was the clearest explanation (avg. 3.74) of the route when showing the map in 13-a. These expressions included all information regarding Starting Point(X), Route(Y), and Destination(Z). Instead of describing complex alleys, the route chosen explains the main streets, selecting landmarks along the route such as Hareza Tower (ハレザタワー), the elevated highway, and Nitori (ニトリ) as reference points.

5 Conclusions

This study aimed to collect clear route information reference expressions using both specific and relative location information for language-based direction instructions to autonomous driving systems. Through crowdsourcing, expressions describing route information on maps were gathered. Each expression was manually annotated to determine whether it contained information regarding the starting point, route, and destination. The data revealed a tendency for the starting point information to be missing from expressions. Additionally, clarity ratings were collected through crowdsourcing, indicating the importance of including information regarding routes and destinations.

Existing studies, particularly map task corpora, have focused on analyzing interactions involving the transmission of new and old information based on map tasks; they lack attention on clear route information explanations using maps. This study, however, uses maps of real locations instead of virtual environments. Generating clearer route information reference expressions from map information requires collecting diverse expressions along with their clarity ratings, and examining what contributes to clarity. This study differs from previous research because it collected route information reference expressions.

In future research, we plan to annotate reference expression information. Relative reference expressions are inherently based on information from three or more points and are known to be abstracted as according to the double-cross model (Freksa, 1992). However, utterances that contain information from three or more points cannot be obtained (Kawabata et al., 2023). Data collected in this study include information from three or more points. Therefore, the data should be annotated based on the double-cross model.

The data gathered in this study, including the map information, classifications, and clarity ratings, are accessible to the public at `https://github.com/masayu-a/HRI-JP-RIRE-DB`.

Acknowledgments

We would like to express out gratitude for the support and collaboration provided by the National Institute for Japanese Language and Linguistics (NINJAL) in the joint research project, "Evidence-based Theoretical and Typological Linguistics." Additionally, we are deeply thankful for the joint research partnership between the Honda Research Institute and NINJAL. This research is partially supported by JSPS KAKEN JP22K13108 and JP19K13195.

References

Michael Barclay and Antony Galton. 2008. A scene corpus for training and testing spatial communication systems. In *AISB 2008 convention communication, interaction and social intelligence*, volume 1, page 26.

Herbert H Clark and Deanna Wilkes-Gibbs. 1986. Referring as a collaborative process. *Cognition*, 22(1):1–39.

Christian Freksa. 1992. Using orientation information for qualitative spatial reasoning. In *Theories and methods of spatio-temporal reasoning in geographic space*, pages 162–178. Springer.

Yoshiko Kawabata, Mai Omura, Masayuki Asahara, and Johane Takeuchi. 2023. Spatial information annotation based on the double cross model. In *Proceedings of the 37th Pacific Asia Conference on Language, Information and Computation*, pages 137–144, Hong Kong, China. Association for Computational Linguistics.

George Lakoff. 1987. *Women, fire, and dangerous things: What categories reveal about the mind*. University of Chicago press.

Stephen C Levinson. 2004. *Space in Language and Cognition: Explorations in Cognitive Diversity*. Cambridge University Press.

George Psathas and Martin Kozloff. 1976. The structure of directions. *Semiotica*, 17(2):111–130.

Amy L Shelton and Timothy P McNamara. 2004. Orientation and perspective dependence in route and survey learning. *Journal of Experimental Psychology: Learning, Memory, and Cognition*, 30(1):158–170.

Holly A. Taylor and Barbara Tversky. 1992. Spatial mental models derived from survey and route descriptions. *Journal of Memory and Language*, 31(2):261–292.

6 Appendix

[Please describe the route from the marker "■" to the marker "★" on the map using surrounding landmarks.]

- Provide an expression of 30 characters or more and less than 200 characters.
- Describe using the perspective of you initially being at the ■ mark on the map, considering front, back, left, and right.
- Refrain from using up, down, left, and right in reference to the map layout.
- Avoid using north, south, east, and west.

Figure 2: Example of the survey screen: Collection of route information reference expressions

[Please assess the clarity of the document describing the route from the marker to the destination, after viewing the map in the following link.]

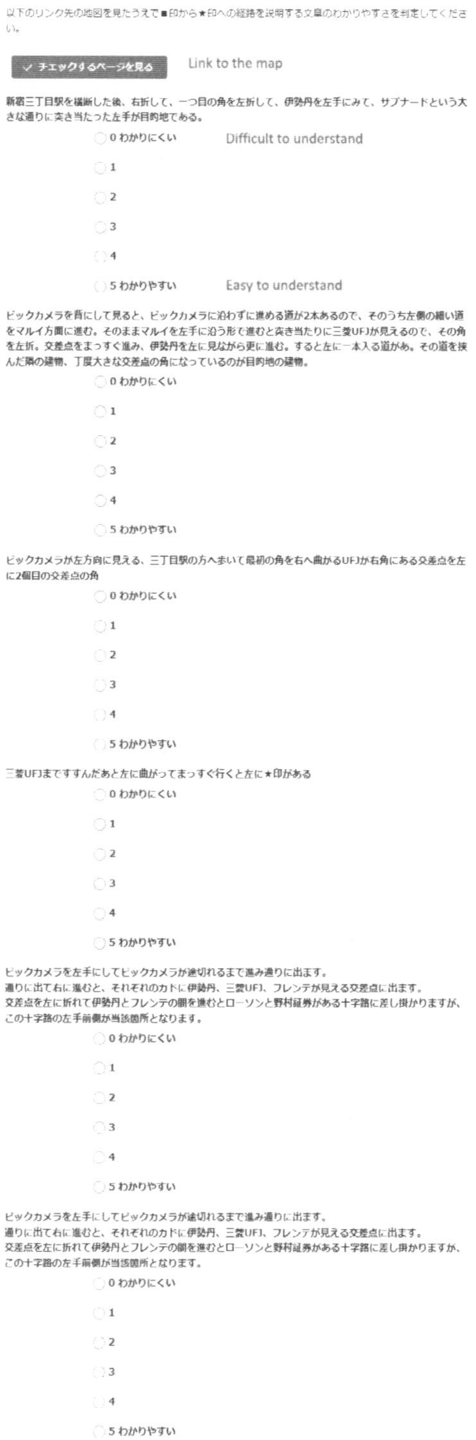

Figure 3: Example of Survey Screen: Clarity Rating Survey

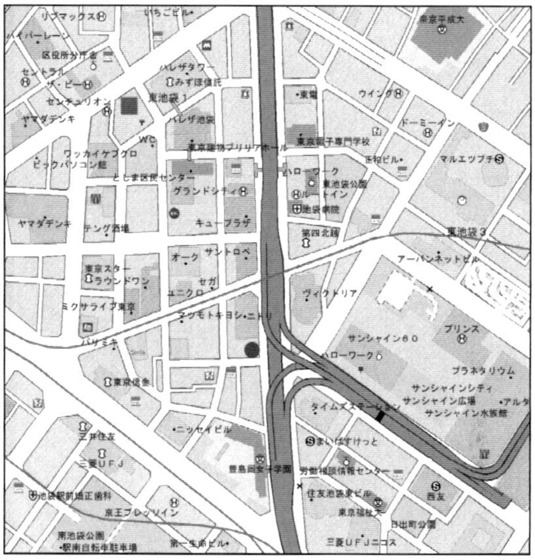

(Example) 大通りをハレザタワー方面にしばらく進むと高速が高架になっている大通りにでます。その大通りを渡らずに右折してください。高架に沿ってしばらく進むと右手にニトリが見えてきます。ニトリを過ぎてすぐ右手に隣接しているのが目的地です。

(Label) W=FALSE, X=TRUE, Y=TRUE, Z=TRUE, Clarity=3.74

(English Translation) Proceed along the main street towards Hareza Tower (ハレザタワー) for a while until you reach a main street with an elevated highway. Turn right without crossing that main street. Continue along the elevated highway for a while, and you will see Nitori (ニトリ) on your right. The destination is immediately adjacent to Nitori(ニトリ) on the right.

Figure 4: Example of collected of route information reference expressions (13-a)

Author Index

Asahara, Masayuki, 30

Hakimov, Sherzod, 17

Kawabata, Yoshiko, 30
Konishi, Hikari, 30

Lin, Albert, 1

Narasimhan, Karthik R, 1
Nguyen, Khanh Xuan, 1

Omura, Mai, 30

Sadler, Philipp, 17
Schlangen, David, 17

Takeuchi, Johane, 30
Tuyls, Jens, 1

Zhang, Alex L, 1